An Inquiry-Based Introduction to Engineering

Michelle Blum

An Inquiry-Based
Introduction to Engineering

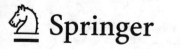
Springer

Michelle Blum
Syracuse University
Syracuse, NY, USA

This book includes lecturer materials at sn.pub/lecturer-material

ISBN 978-3-030-91473-8   ISBN 978-3-030-91471-4   (eBook)
https://doi.org/10.1007/978-3-030-91471-4

This Springer imprint is published by the registered company Springer Nature Switzerland AG
The registered company address is: Gewerbestrasse 11, 6330 Cham, Switzerland

For Thomas, Mackenzie, and Andrew—
forever my priorities.

Preface

This Introduction to Engineering textbook is created using Inquiry-Based Learning (IBL) methodology. Inquiry-based approaches are forms of inductive teaching that emphasize the student's role in the learning process [1]. During class time, students are empowered to explore concepts, ask questions, and share ideas. Instead of the instructor lecturing to passive listeners about the material. A primary aim of IBL strategies is student involvement, which leads to increased understanding and retention of material. All IBL strategies use common elements including active small-group discussion [2].

Inquiry-based learning models are well established in education. Originating in the 1960s, the inductive pedagogies have been used in primary and secondary school education [3–5]. There are several different inquiry-based instruction levels with various names and nomenclature depending on the discipline [6, 7].

- Level 1 (conformation inquiry) is where students follow the lead of a teacher as the entire class engages to confirm a principle or concept through an activity, where the results may or may not be known in advance. An example of this would be the whole class working on an experiment with the teacher working through the steps with the students.
- Level 2 (structured inquiry) is when students investigate a teacher presented question using teacher prescribe procedures. An example of this is students working on a teacher made worksheet while the teacher walks around the class to answer questions.
- Level 3 (guided inquiry) is where students investigate a teacher-chosen topic/question using student designed or selected procedures. For example, students designing an experiment to solve a specific problem a professor gives them.
- Level 4 (open inquiry) is where students choose their own open topics or questions to investigate, and they use procedures that are student formulated. An example is students choose and design their own product to solve a world-wide need.

In primary education, K-12 teachers are encouraged to scaffold the levels into their curriculum, beginning with Level 1 and working up to Level 4. Beginning instruction at the lower levels help to develop student's inquiry skills and their motivation and excitement for the learning methods [8]. However, in secondary engineering education, IBL is rarely implemented comprehensively throughout a class, and it is highly subject matter dependent. Particularly, STEM disciplines (science, technology, engineering and math) trend toward the use of Levels 1 through 3 for their required undergraduate classes, with may be the exception of a senior capstone class, while other disciplines (social science, language arts, and education) are suited to more prevalently utilize Level 4 [9]. The effectiveness of IBL has been assessed at a range of college institutions and for a variety of courses. At the college level, IBL methods have been used for teaching general chemistry [10], foreign language [11], information technology [12], computer science [13], and materials science [14].

Currently, textbooks are only on the market for freshman level math and science courses [15, 16], and sophomore or junior level engineering science courses [17, 18]. If students are not exposed to this type of learning early on, they tend to be resistant in supplemental classes. Therefore, this textbook aims to expose students to IBL instruction in Introduction to Engineering classes which will increase success of student learning with IBL at any level of engineering education.

The content of the first edition of the book includes three parts, with each part consisting of four to ten chapters. The first section of the book focuses on the knowledge and strategies for being an effective student. The second section of the book focuses on the Engineering Profession, and the final section focuses on technical aspects of engineering fundamentals. Details of each chapter can be found in the Table of Contents.

Each topic begins with a few brief paragraphs of introductory and fundamental material. Followed by a series of focused questions that guide the students' learning about the concept(s) that are being taught. All four IBL levels are utilized in the book, with three of the four general types (confirmation, structured, and guided inquiry) mainly being used in the technical chapters and open inquiry being incorporated into the student learning and engineering profession sections. A typical class period is organized and executed as follows:

1. When a new topic is introduced, a brief introduction is given. Alternatively, a review of the previous classes' material was given if it were a continuation of topic.
2. Students are released to work in groups of 3–4 on a specified set of topic questions.
3. The groups work on the activity for a designated amount of time. The amount of time varies depending on the level of the IBL activity and the components of the worksheet. Typically, the minimum time given for a worksheet is 7–8 min and the maximum is 20 min.
4. While the students were working, the instructor walks around and observes each group. When students ask questions, the instructor should try not to directly answer the worksheet questions, but rather guide the students to discover the answer themselves.

5. If the instructor observes that most of the class is struggling with a particular section or question, they should stop the group work briefly and address the entire classes' confusion.

6. Once the allotted work time was up, the instructor can actively engage the class in reviewing the answers to the worksheet. Keep in mind that the instructor should not directly give the answers. The class is responsible for explaining the answers along with the instructor.

7. Steps 1 through 6 are repeated for each topic. For a 50-min class period, typically 2–4 topic sets could be covered and for an 80-min period 3–5 topic sets can be completed. Students can either utilize the textbook (if in paperback form) as a workbook, or take notes separate. The topic sets act as class notes for studying, homework to be turned in, and reference for completing practice problems.

The use of the IBL methods in this text will greatly improve the teaching and learning of first year engineers. It will educate students early in their college career to the benefits and skills essential to inductive learning. The supplemental materials are provided for the instructor with the goal of helping to ease unfamiliarity and increase comfort with teaching IBL methods and to streamline course preparation. This textbook is the start to IBL learning for engineering students who overtime will see improved satisfaction in their learning and retention in the discipline.

Syracuse, NY, USA   Michelle Blum

References

1. Prince, M. J.; Felder, R. M. "Inductive Teaching and Learning Methods: Definitions, Comparisons, and Research Bases", Journal of Engineering Education 2006, 95, 123–138.
2. Prince, M., and R. Felder. 2007. The many faces of inductive teaching and learning. Journal of College Science Teaching 36, no. 5: 14–20.
3. Schwab, J. (1960) Inquiry, the Science Teacher, and the Educator. The School Review 1960 The University of Chicago Press.
4. Maaß, K., Artigue, M. Implementation of inquiry-based learning in day-to-day teaching: a synthesis. ZDM Mathematics Education 45, 779–795 (2013). https://doi.org/10.1007/s11858-013-0528-0
5. Witt, C, The Impact of Inquiry-Based Learning on the Academic Achievement of Middle School Students, 2010 Western AAAE Research Conference Proceedings, 269–281 (2010).
6. Banchi, H. & Bell, R. (2008). The Many Levels of Inquiry. Science and Children, 46(2), 26–29.
7. Inquiry Mindset, Trevor MacKenzie, Rebecca Bathurst-Hunt, 2018, ElevateBooksEdu.

8. Trevor Mackenzie book, Yoon, H., Joung, Y. J., Kim, M. (2012). The challenges of science inquiry teaching for pre-service teachers in elementary classrooms: Difficulties on and under the scene. Research in Science & Technological Education, 42(3), 589–608.

9. Vajoczki, Susan; Watt, Susan; Vine, Michelle M.; and Liao, Rose (2011) "Inquiry Learning: Level, Discipline, Class Size, What Matters?" International Journal for the Scholarship of Teaching and Learning: Vol. 5: No. 1, Article 10.

10. John J. Farrell, Richard S. Moog, and James N. Spencer, A Guided Inquiry General Chemistry Course, Journal of Chemical Education, Vol. 76 No. 4 April 1999.

11. Johnson, C. (2011). Activities using process-oriented guided inquiry learning (POGIL) in the foreign language classroom. Die Unterrichtspraxis 44(1), 30-IV. Retrieved from http://search.proquest.com.ezproxy.rit.edu/docvie w/878895308?accountid=108.

12. Myers, T., Monypenny, R. & Trevathan, J. (2012). Overcoming the glassy-eyed nod: An application of process-oriented guided inquiry learning techniques in information technology. Journal of Learning Design, 5(1), p. 12–22.

13. Clifton Kussmaul, Process Oriented Guided Inquiry Learning (POGIL) for Computer Science, SIGCSE'12, February 29–March 3, 2012, Raleigh, North Carolina, USA.

14. Douglas, E.P. (2014) Materials Science and Engineering: A Guided Inquiry, Upper Saddle River, NJ, Pearson.

15. R. S. Moog and J. J. Farrell, Chemistry, a guided inquiry, 2nd edition, John Wiley & Sons, New York, 2002.

16. A. Straumanis, C. Beneteau, Z. Guadarrama, J. E. Guerra and L. Lenz, Calculus I: A guided inquiry, Wiley, Hoboken, NJ, 2014.

17. E. P. Douglas, Materials science and engineering: A guided inquiry, Pearson Education, Upper Saddle River, NJ, 2014.

18. J. N. Spencer, R. S. Moog and J. J. Farrell, Physical chemistry: A guided inquiry. Thermodynamics, Houghton Mifflin Company, Boston, 2004.

Acknowledgments

All of the images in this textbook were expertly created by Ms. Bailey Rose Kretschmer.

Contents

About the Author

Michelle Blum earned dual B.S. degrees in Physics and Mechanical Engineering from the State University at Albany and Rensselaer Polytechnic Institute before going on to earn her masters and Ph.D. degrees from the University of Notre Dame. She is currently an Associate Teaching Professor, Director of the Mechanical Engineering Undergraduate Program and Dean's Faculty Fellow for Assessment at the highly rated R1 institution, Syracuse University in New York. She has focused on teaching and engineering education for almost a decade and has taught multiple classes across mechanical and aerospace engineering undergraduate curriculums, ranging across all levels. She is deeply passionate about providing students with educational material that enables them to learn by actively engaging with course content, as well as creating streamlined materials for instructors. For her teaching, she has received numerous awards including the 2016 Syracuse University Laura J. and L. Douglas Meredith Professorship Teaching Award, 2017 TACNY College Educator of the Year, 2017 SU Dean's Award for Excellence in Engineering Education, and the 2018 The Filtertech, Pi Tau Sigma, and Sigma Gamma Tau Award for Excellence in Education in Mechanical and Aerospace Engineering. She has been serving as an ABET Program Evaluator for ASME, evaluating mechanical engineering B.S. programs since 2021. For her assessment work the Mechanical Engineering B.S. program at Syracuse earned a One University Assessment Award for Outstanding Assessment in 2022.

Part I
Being an Effective Student

Chapter 1
Achieving Success in Engineering

Abstract This chapter introduces fundamental strategies engineering students can use to be successful throughout their engineering curriculum. This chapter begins by discussing and reflecting on what success is based on everyone's objectives. Then we work to focus those objectives on personal goals. Then we work through approaches to strengthen your commitment and develop strategies to achieve success by working harder, smarter, and thinking positively. We also work to define ourselves, who we want to be as people and as engineers. Finally, we work to develop an outline for your time in your degree program and a structured plan to accomplish all your goals.

By the end of this chapter, student will learn to:

- Reflect on what success means to you and focus those objectives on personal goals.
- Explain different mindsets on ability and effort and reflect on your mindset.
- Explain different ways students are motivated and reflect on what motivates you.
- Explain the importance of effort over ability and begin to develop strategies for increasing work effort.
- Explore personal values and approaches that will be employed throughout your college career.
- Explain extracurricular and student involvement opportunities that are available throughout your college career.
- Identify what certain opportunities you might want to pursue through college.
- Identify information about institutional resources available to help you throughout your undergraduate career.
- Plan your path to graduation.

© Springer Nature Switzerland AG 2022
M. Blum, *An Inquiry-Based Introduction to Engineering*,
https://doi.org/10.1007/978-3-030-91471-4_1

1.1 Setting Goals

You are in this introduction to engineering class because you were accepted into the engineering college at your prestigious institution. You are about to embark on an incredible journey of maturation, academic, and self-discovery. There will be many obstacles on the path to success but understand that *success is a process*. And learning that process will increase your chances for achieving success. The initial step, and one of the biggest hurdles to success, is *defining* it. The dictionary defines **success** as achieving something that is desired, planned, or attempted [1]. Without clearly defining *what is desired* and *how you plan to achieve it*, there is no success. Therefore, the first step to success is identifying a clear set of goals.

Defining a clear set of goals gives a person focus, control over their life's direction, and it provides a benchmark for assessing and determining whether they are being achieved. Following a specific process for goal setting helps people formulate accomplishable goals. There are a variety of suggestions for goal setting processes, but the process described below focuses on five key steps [2]

1. **Set goals that motivate**. Make sure that the goals being set are ones that are important to you. This will motivate you and strengthen your commitment to accomplishing them. Set goals related to the high priorities in your life and with each goal include *WHY* it is important and valuable.
2. **Set S.M.A.R.T Goals**. S.M.A.R.T goals were defined by business consultant George Doran in 1981 [3], as a way of establishing goals that have a meaningful impact. There are many variations of what S.M.A.R.T stands for, but fundamentally goals should be
    - Specific —Setting specific goals mean that they are well defined. Name precisely where you plan to end up.
    - Measurable—The goals should include precise amounts, a timeline, and dates so that the degree of success can be measured. For example, if you want to secure an internship for the summer a timeline must be set for things like when to get your resume completed and a quantifiable number must be set for how many resumes will be sent to companies each week.
    - Attainable—Set goals that are realistic yet challenging. If the goal is too hard, it will not be accomplished and it will dishearten you. If the goal is too easy, then there is no real work involved and thus no real personal growth.
    - Relevant—Set goals that are relevant to the direction you want your career and life to take. Aligning goals will keep your time and energy focused.
    - Time Bound—Give all your goals a deadline. Setting a date and timeframe for accomplishment will help you achieve goals in an efficient manner and allow time to reflect and celebrate once they are achieved.
3. **Write goals down**. Setting the goals in writing makes them tangible and harder to forget. This can take many forms such as a "To-do" list or "Goals" list.
4. **Create an Action Plan**. Write out the individual tasks (with timeline included) that need to be completed to accomplish the goals. This is the hard work that needs to be applied to succeed. You can also check the tasks as you complete them. This will show the progress toward achieving the goal.

5. **Never give up!** Whether short- or long-term goals you need to develop a "stick with it" attitude. Constantly remind yourself to review your goals and action plan. This will help you stay on track and committed to finishing.

This activity helps to begin setting goals for you throughout your undergraduate engineering career which will lead to academic professional and personal success.

1.1.1. **List one academic goal you want to set for yourself. (Remember to use the S.M.A.R.T. paradigm)**

1.1.2. **Explain why this academic goal is important to you.**

1.1.3. **List at least three tasks (it can be more) that are necessary to achieve this academic goal.**

1.1.4. **What date/timeframe do you want to have this academic goal accomplished by?**

1.1.5. **List one behavioral or social goal you want to set for yourself. (Remember to use the S.M.A.R.T. paradigm)**

1.1.6. **Explain why this social goal is important to you.**

1.1.7. **List at least three tasks (it can be more) that are necessary to achieve this social goal.**

1.1.8. **What date/timeframe do you want to have this social goal accomplished by?**

1.1.9. **List one personal goal you want to set for yourself. (Remember to use the S.M.A.R.T. paradigm)**

1.1.10. **Explain why this personal goal is important to you.**

1.1.11. **List at least three tasks (it can be more) that are necessary to achieve this personal goal.**

1.1.12. **What date/timeframe do you want to have this personal goal accomplished by?**

1.2 Changing Attitude

When pursuing a degree in the rigorous discipline of engineering having a positive mindset is very important for success. Understanding the relationship between effort and ability can keep you optimistic and motivated, even during difficult moments. **Ability** is defined as possession of the means or skill to do something [1]. **Effort** is defined as the attempt to do or achieve something. American psychologist Carol Dweck [4] explained that people have two main mindsets about success based on the underlying nature of ability versus effort. The general mindsets are called **growth mindset** and **fixed mindset**. Students with **growth mindset** think that intelligence and abilities can be developed through effort, persistence, trying different strategies, and learning from mistakes. Students with a **fixed mindset** believe that a person's intelligence and abilities are fixed traits; something that someone is born with and that cannot really change.

Similar to mindset, research on achievement motivation studies how students think about themselves, their tasks, and their performance [5]. There are many theories about students' motivation, but the main two ideas are that students are either

mastery-oriented or **performance-oriented**. Students who are **performance-oriented** are motivated to achieve based on external factors such as grades, scholarships, awards, or accolades from other people. Their sense of achievement is heavily entwined with the grades they earn. Students who are **mastery-oriented** focus on learning material and filling their knowledge gaps. They are driven by internal factors such as increasing their understanding and competency at a task or concept. Different factors drive different people, and it is good to understand why you are motivated to perform well.

Finally, here are some helpful tips about focusing your attitude:

- *Do what you love, even when it is difficult*. Being passionate about your work improves its quality and your satisfaction.
- *Make sure whatever you do is in the service of others*. Serving others instead of yourself improves your awareness, the impact you have, and the response you receive from others.
- *Take responsibility for all aspects of your learning and your work*. If you do not understand something, reread, restudy, or ask for help. This will remove any obstacle that otherwise would impede your success.

This next set of questions helps us to reflect on attitude.

1.2.1. **Given the following scenarios, tell whether the student has a growth mindset or fixed mindset**
- **Molly is a sophomore mechanical engineering student who is struggling in statics, she feels she has no control over her abilities, and is helpless to improve. She is disheartened that engineering ended up being so difficult. Since there is no point in trying anymore, she is going to withdraw from the class.**
- **Ari is a sophomore who is also struggling in statics. He believes that he can improve his performance if he practices more example problems and attends the professor's office hours more frequently. He does not feel threatened by the hard work and he is not afraid to ask his friends for help on assignments he does not understand. And although he failed the first exam, he knows there are three more exams so he can improve his grade.**

1.2.2. **Based on the following fixed mindset phrases in Table 1.1, fill in a new growth mindset phrase to replace it.**

1.2.3. **Tell about your high school performance. Where did you excel? Where did you struggle? Explain what motivated you to make choices in high school.**

1.2.4. **Explain what motivated you to become an engineer? Where do you expect to excel? Where do you anticipate you will struggle?**

1.2.5. **Which achievement motivation do you think would be best to use as an engineering student? Explain why.**

Table 1.1 Common fixed mindset statements

Fixed mindset statement	Growth mindset statement
I cannot do this	
I give up	
This is too hard	
My plan did not work	
It is good enough	
This is not my strength	
I cannot make this any better	
He/She is so smart. I will never be as smart	
I cannot do math	

1.3 Changing Effort

Academic success in studying engineering is predominantly under the student's control. The amount of effort a student puts forth is directly correlated to their success. Generally, insufficient efforts lead to poor academic performance. **Effort** is defined as a vigorous or determined attempt [1]. The effort you devote to your studies consists of two key components—your time and your energy. To complete certain tasks, such as a homework or project assignment, requires a student to commit to spending time and mental energy (focus). Subsequent chapters will discuss details about how to make the most of your time and energy so that your effort can be efficient and effective. However, here are some suggestions on how to increase your work effort:

- Break big goals into small tasks.
- Start your day with the most important task.
- Surround yourself with motivated people.
- Limit distractions so you stay focused.
- Give yourself breaks to refocus your efforts.
- Remember your "why."
- Reward yourself when you have completed major milestones.
- Take care of yourself physically.

This next activity lets us reflect on our current level of effort and explore strategies to increase effort.

1.3.1. **In high school, how much time each day did you spend on schoolwork?**
1.3.2. **Did you employ any of the strategies above when completing your schoolwork? If so, which ones and how often?**
1.3.3. **In high school, how much time each day did you spend on extracurricular work or activities? What were they?**

1.3.4. **Did you employ any of the strategies above when completing your extra-curricular work or activities? If so, which ones and how often?**
1.3.5. **Write down three effort strategies you plan to keep the same from high school.**
1.3.6. **Write down three new effort strategies you plan to adopt and employ in college.**

1.4   Changing Approach

One definition for **approach** is a way of dealing with something [1]. To be success-ful in your pursuit of an engineering degree it may be necessary to change your approach. What this means is that the attitude and methods you used in high school may need to be altered. *How* you go about your college career and studies is impor-tant. There is a saying "Work smarter, not harder." While this may apply to many areas, in engineering the phase should be "Work smarter AND harder." The previous section discussed working hard, and subsequent chapters will discuss specific tac-tics for working smart. But to define yourself what kind of engineer you want to be, you first need to define what type of person you want to be. This next activity explores our personal values so we can begin to adjust our attitudes and approaches for the future.

1.4.1. **Make a list of personal values. They can be things that you or somebody else might value in life. List as many things as you can.**
1.4.2. **Choose two of the values from your list that are the most important to you. For each value you select describe why that value is important to you.**
1.4.3. **List three ways that your approach in college will help you obtain or maintain the values you listed above. If you need to change your approach, list three ways your approach will change.**

1.5   Extracurricular and Student Involvement Opportunities

Even though earning an engineering degree is a major step to acquiring an exciting and rewarding job, companies value students who have shown not only academic aptitude but additional abilities as well. For a comprehensive college experience, there are many extracurricular and involvement opportunities engineering students should pursue. Real-life and international experiences will lead to increased per-sonal, professional, and leadership proficiency. Below are several common opportu-nities that most universities provide for their engineers to pursue:

- **Student Clubs**. At every university, there are a plethora of activities and clubs for students to become involved with. Students in college engage in a wide vari-ety of activities ranging from engineering societies, performing arts groups, to intramural sports. Typically, there is an office of student activities that manage them and at the beginning of an academic year many times a student activities fair is held to publicize them.

- **Cooperative Education and Internships**. Known as **Co-ops**, students partici-pate in experiential learning by spending alternating time working full-time for a company, then returning to their engineering studies. For example, a student would accept a co-op rotation for a summer and then the fall semester and then return as a full-time student in the spring semester and then return to the com-pany for the following summer. **Internships** are like co-ops in that they offer career-related experience in professional settings however internships are of a more limited timeframe. Typically engineering students go on internships for a specific number of weeks during the summer. Students can participate in multi-ple internships throughout their college career. Internships and Co-op assign-ments have many benefits including they allow students to gain real work experience, they put the theories they learn in the classroom into practice, they develop great professional contacts, and it is a good tool to determine if engi-neering is the right major. Typically, a university will have a career services office, or a separate cooperative education office, which helps students identify co-op and internship opportunities.
- **Research Experience for Undergraduates**. Known as **REU**, most engineering schools have active research programs, with faculty members performing cutting edge experiments. Students participate in research and scholarship with faculty members either at their home university, another, or at a national or government laboratory. There are many benefits for students who participate in REUs. First, it broadens and deepens their classroom learning and supports the development of a range of academic and professional skills, it helps student cultivate an under-standing of research design and methodology, and students develop strong men-toring relationships with faculty members. Many summer REU programs are sponsored by the university and the National Science Foundation (NSF). Students can also perform research during the academic year. Undergraduates can per-form research for academic credit or paid employment.
- **Study Abroad**. This is an opportunity that many universities offer where stu-dents choose to pursue some of their college studies in a foreign country. The benefits to study abroad are that students continue to develop strong technical skills while gaining a global perspective. International experiences help students stand out to prospective employers because employers are increasingly looking for students with cross-cultural skills and an understanding of the world's diver-sity. Students typically spend a quarter or a semester abroad. There are varieties of programs among universities. Exchange programs are where students con-tinue to pay their home institution's tuition but attend a foreign university. Direct enrollment programs are where students directly enroll and pay tuition at an overseas location. There are also faculty-led programs or university centers where faculty of the home institution teach courses aboard. For detailed informa-tion and to find a program of interest, students should seek out their university study abroad office.

This next set of questions helps us begin to research extracurricular and student involvement opportunities that are available and to reflect and identify what opportunities you might want to pursue.

1.5.1. **Perform the following tasks and answer the following questions about student clubs:**
   - **List five clubs, societies, or activities that you are interested in joining or participating in while in college.**
   - **What is the website where your university lists information about student clubs, activities, and organizations? Are there multiple websites? (write down in your notes for future reference)**
   - **Can you find the five you listed above on the student activities website?**
   - **For the ones you can find on the website list any relevant information you find. (For example, contacting information for the club, club website, club informational meeting dates, etc.)**
   - **For the ones you could not find on the website, brainstorm how you could find more information about these interests. For example, if there is no club/activity would you consider starting one up?**

1.5.2. **Perform the following tasks and answer the following questions about Co-ops and Internships:**
   - **Does your institution have a dedicated co-op or career services office that aids with finding and maintaining co-op placement? If so, where is it located and what is the contact information of one person in that office? (write down in your notes for future reference)**
   - **How many months and semesters of co-op experience does your institution support during a student's undergraduate program?**
   - **Research and reflect on your options for internships or co-ops. Would you rather participate in internship or co-op opportunities? Why one over the other?**
   - **When during your undergraduate career would you like to initially participate in an internship or co-op?**

1.5.3. **Perform the following tasks and answer the following questions about undergraduate research:**
   - **Find your university or college's engineering faculty website, look at the different research areas of faculty members in your department. Are there areas of research you would be interested in working in?**
   - **Look at the following website: National Science Foundation's Research Experiences for Undergraduates (REU) program (website: https://www.nsf.gov/funding/pgm_summ.jsp?pims_id=5517).   Are there any programs that you would be interested in attending?**
   - **Think about your engineering curriculum. List when in your curriculum you would like to perform undergraduate research (it could include during the semester or over the summer).**

1.5.4. **Perform the following tasks and answer the following questions about study abroad:**
- **List 1–3 countries where you would be interested in studying abroad.**
- **Does your institution have a dedicated study abroad services office that aids with finding and coordinating study abroad opportunities? If so, where is it located and what is the contact information of one person in that office. Is there a website? (write down in your notes for future reference)**
- **For your institution and program, when are applications for study abroad due?**
- **For your institution and program, how much does it cost to study abroad and what is the payment structure?**
- **For your institution and program, what study abroad credits work with your degree program?**

1.6 Information to Know

Throughout your undergraduate career pursuing engineering you will have many questions. So it is important to know where to go and who to ask for help. One of the major reasons students struggle during their first few semesters in college is that they do not ask for help. For questions pertaining to your technical classes, you should always ask the professor teaching the course. But for other more nebulous questions, it is important to identify the resources on your campus. This next set of questions helps in identifying helpful intuitional resources.

1.6.1. **Every incoming student is given a college identification card. Look up your student ID number. You do not have to write it down but begin to memorize it. It is how you will be identified by the university. Also, the card most likely provides access to many resources at your institution. Find a safe location for it.**
1.6.2. **What is the semester you are incoming as a first-year student? What is the semester you are expected to graduate?**
1.6.3. **College and/or Department personnel: Fill out the following information so that you will have the names and contact information of people in the department and college to ask for help.**
- **Director of your undergraduate engineering program: Name, Office, Phone number, Email**
- **Chair of your Engineering department: Name, Office, Phone number, Email**
- **Dean of your College of Engineering: Name, Office, Phone number, Email**

1.6.4. **Other resources. From the list below, identify the person's name, office, phone number, and email of who you would contact for the following resources:**
- **Academic advising**
- **Career services**
- **Student records (AP and transfer credits, etc.)**
- **Counseling/Health and Wellness services**
- **Title IV and Ombudsperson Office**
- **Student services**

1.7   Develop a Plan

It is a good idea to sit down and plan your graduation. There are many options and paths to take, and everyone will have a unique experience. Think about your current interests and the previous sections. If you are unsure about your interests in a section, that is okay, you can investigate the area more before committing to anything. Remember, it is okay if your interests change, you can always modify your plan later. Continually revisit the plan and discuss with interested parties such as your family, friends, and academic advisors.

1.7.1. **Obtain your program curriculum sheet (it might also be found in an online course catalog). Using the program schedule, write in an academic plan to graduation. Include things such as:**
- **Require degree classes**
- **Electives**
- **Minor classes**
- **What classes (if any) are replaced by AP, IB, or transfer credits**
- **Summer plans such as internship, co-op, research, or study abroad**
- **Any semesters that will be replaced by internship, co-op, or study abroad**

End of Chapter Questions

IBL Questions

IBL1: Answer the following questions and then discuss the answers with a partner or your team.

1. Establish a goal for a grade for each class you are taking this semester.
2. If you achieve each grade, calculate the GPA you would earn.
3. List the tasks necessary to achieve each grade.
4. Develop a timeline for the tasks previously listed.
5. Explain why achieving each grade is important to you.

IBL2: Answer the following questions and then discuss the answers with a partner or your team.

1. Do you consider yourself a fixed mindset learner or a growth mindset learner? Explain why.
2. List three practices you can adopt to become a stronger growth mindset learner.
3. List three techniques you can use if your friend is a fixed mindset learner.
4. Do you consider yourself a mastery-oriented learner or a performance learner? Explain why.
5. List three practices you can adopt to become a stronger mastery-oriented learner.
6. List three techniques you can use if your friend is a mastery-oriented learner.

Practice Problems

1. Write an essay describing the job you envision yourself having in 10 years. Discuss the education level(s) and skills necessary for the job and explain how to plan to obtain them.
2. Explain the steps necessary to declare a minor or concentration at your institution.
3. Write an essay describing a study abroad experience you would like to have. Explain the steps that need to be taken by yourself and your institution in order for you to have that experience.
4. Write an essay describing a cooperative education or internship experience you would like to have. Explain the steps that need to be taken by yourself and your institution in order for you to have that experience.
5. Write an essay describing a research experience you would like to have. Discuss the area of expertise you are interested in studying and if there are any faculty who perform research in that area at your institution. Explain the steps that need to be taken by yourself and your institution for you to have that experience.

References

1. Oxford English Dictionary. Oed.com.
2. Golden Rules of Goal Setting, mindtools.com, viewed May 12, 2021.
3. Doran, G. T. (1981). "There's a S.M.A.R.T. way to write management's goals and objectives". Management Review. 70 (11): 35–36.
4. Dweck, Carol S (2006) Mindset: The New Psychology of Success, Random House, ISBN: 9780345472328.
5. Ames, Carole (1992). "Classrooms: Goals, structures, and student motivation". Journal of Educational Psychology. 84 (3): 261–271. doi:https://doi.org/10.1037/0022-0663.84.3.261. ISSN 1939-2176.

Chapter 2
Teaching and Learning Approaches

Abstract Most students who pursue engineering degrees are typically top performers in high school, so most likely they never needed to reflect on how they learn or process information from their teachers. However, to be a success in college engineering courses understanding teaching and learning processes is important. To be academically successful students need information about their personal learning styles in conjunction with their professor's teaching styles to employ tailored study strategies. To begin this chapter the teaching process is discussed. A range of teaching styles are explained and the benefits of understanding how a professor teaches are examined. Next, the learning process is explored. The variety of ways a student receives and processes information is defined. The opportunity is also given for reflection to find insights into your own preferred learning processes and to plan how to leverage that knowledge for academic success. Finally, common mistakes students make that hinder their learning are reviewed.

By the end of this chapter, student will learn to:

- Define different methods and styles of teaching.
- Explore your preferred teaching styles and identify strategies for competency of learning.
- Describe the learning process and define its components.
- Discover your preferred way of learning.
- Explain steps to improving your learning process.
- Develop strategies to overcome common student learning mistakes.

2.1 Teaching Approaches

Teaching approach refers to the general principles, pedagogy, and management strategies used for classroom instruction [1]. The teaching approach generally consists of the *teaching method*, *teaching style, and instruction elements*. Many times, the teaching approach of a professor's classroom depends on the education philosophy of the professor and the subject area being taught. It is important to identify the teaching mode of a class because this knowledge will guide you and let you plan your learning process for the course material.

There are a variety of teaching *styles* used in education [1]. They can be organized into four categories based on two major considerations.

- *Consideration 1: Who is responsible for the learning?*

  - The **teacher-centered** style means the teacher is prominently responsible for students learning. Students are considered "blank slates" who passively receive knowledge from their teacher.
  - The **student-centered** style means the students are prominently responsible for their learning. The teacher plays more of a facilitator role while the students build upon past knowledge to acquire new concepts and skills.

- *Consideration 2: What technology facilitates learning?*

  - A **high-tech** teaching style is one that relies mainly on different virtual resources to enable learning.
  - A **low-tech** teaching style is one that relies mainly on more physical resources to enable the learning.

The *method* of teaching refers to how an instructor governs their classroom and implements instruction. It is a combination of the assessment and instructional practices [1]. There is a vast number of teaching methods but below is a list of common teaching methods found in engineering courses.

- **Direct Lecture**—teaching strategy that relies on teaching through lectures and teacher-led demonstrations while students passively listen, watch, and maybe take notes.
- **Hands-on Instruction**—teaching strategy where students perform physical activities requiring students to do, make, or create to learn.
- **Differentiated Instruction**—teaching strategy where instructors tailor instruction or assessment to meet different student needs, such as student access to content based on their performance.
- **Inquiry-based Learning**—teaching strategy based on students answering questions. The questions could be created by a teacher or developed by the students. Students learn by building upon their prior knowledge, researching the questions, finding information and sources, and then explaining key concepts and solutions to problems.
- **Game-Based Learning**—teaching strategy where students work on quests to accomplish a specific learning goal. The students choose actions and complete tasks along the way. As students make certain progress or achievements, they can earn awards, such as badges or points, just like in a video game or board game.
- **Flipped Classroom**—teaching strategy where students watch prerecorded lessons at home and complete in-class assignments.

The *instruction elements* refer to the day-to-day components of the course. Dr. Richard Felder, a recognized expert in the field of engineering education, revealed that there are nominally five instructional elements, each with two options, for the approach in a class [2]. They are listed below

- *What is the type of knowledge being taught?*

  - **Concrete** knowledge deals with facts, data, and observable phenomena
  - **Abstract** knowledge deals with principles, concepts, theories, and mathematical models

- *What mode is the knowledge presenting in?*

  - **Verbal** mode uses mainly spoken and written word
  - **Visual** mode uses pictures, diagrams, films, demonstrations

- *How is the presentation of knowledge organized?*

  - **Deductive** organization starts with communicating the fundamentals of the proceeding to applications
  - **Inductive** organization starts with sharing applications and then progressing to the fundamentals

- *How do students participate?*

  - **Active** participation involves students talking, reflecting, and solving problems
  - **Passive** participation involves students watching and listening

- *What viewpoint does the knowledge come from?*

  - **Global** perspective is when the context and relevance of the knowledge are provided
  - **Sequential** perspective is when there is a step-by-step progression for the knowledge

Not every professor will teach the way you want, but it is important to remember you can be competent in learning even if it is not your preferred teaching method, style, or elements. This next set of questions will let us define different teaching approaches, explore your preferred teaching styles, and identify strategies for increasing our competency of learning. Then the next section will help us uncover our preferred methods for learning.

2.1.1. **In Fig. 2.1 what is the *x*-axis mapping?**
2.1.2. **In Fig. 2.1 explain what increases along the *x*-axis?**
2.1.3. **In Fig. 2.1 what is the *y*-axis mapping?**
2.1.4. **In Fig. 2.1 explain what increases along the *y*-axis?**
2.1.5. **Look at Fig. 2.1 and label the various methods in terms of the four teaching style considerations and quantify each as heavy, medium, or light. The first one is completed for you.**

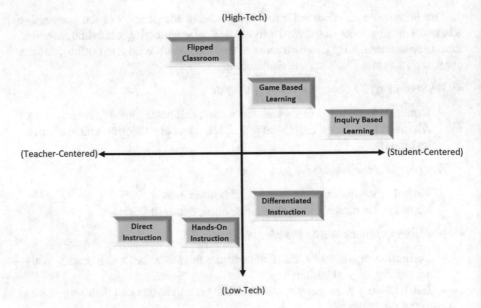

Fig. 2.1 Graph displaying the various teaching methods in terms of the teaching considerations technology and who is responsible for learning

- **Direct Lecture—heavy teacher-centered; heavy low-tech**
- **Hands-on Instruction**
- **Differentiated Instruction**
- **Inquiry-based Learning**
- **Game-based Learning**
- **Flipped Classroom**

2.1.6. **Look at Fig. 2.1 and the list above and think about previous classes you have taken (or are currently taking) in terms of the approaches above. Which approaches have you seen? Which approaches did you like? Which approaches did you dislike? Explain why.**

2.1.7. **Look at Fig. 2.1 and the list above and think about previous classes you have taken (or are currently taking) in terms of the methods above. Which methods have you seen? Which methods did you like? Which methods did you dislike? Explain why.**

2.1.8. **Look at Fig. 2.1 and the list above and think about previous classes you have taken (or are currently taking) in terms of the instruction elements above. Which elements have you seen? Which elements did you like? Which elements did you dislike? Explain why.**

2.1.9. **If you are having difficulty in class because of the way a professor teaches, is that an excuse to not try in the class?**

2.1.10. **List three strategies you can employ if you are having difficulty in class because of the way a professor teaches.**

2.2 Learning Approaches

Learning is defined as the process of acquiring something new [1]. There are many kinds of learning such as cognitive, affective, and psychomotor learning. Psychomotor learning is acquiring new physical skills such as movement, coordination, dexterity, strength, and speed. Psychomotor learning will be happening during your engineering degree as you develop your fine motor skills performing tasks such as sketching, machine tool operations, and working in laboratory class. Affective learning is acquiring new attitudes, values, or emotional responses. Affective learning occurs throughout your academic career (including working through Part I of this book!) as you mature and continually reflect on and strengthen your attitude, awareness, empathy, integrity, and responsibility. Cognitive learning is acquiring new knowledge or intellectual skills. This is the main kind of learning that occurs within your engineering classes.

Each class will be structured with learning objectives, assessments, and activities to increase cognitive learning based on **Bloom's Taxonomy** [3]. Named after educational psychologist Benjamin Bloom, this is a list of six increasing levels of intellectual skills that cognitive learners should strive to achieve. A revised set is shown in Fig. 2.2 [4] and will be used throughout your engineering curriculum by professors to structure class and design coursework. Recognizing these taxonomies will help you identify the purpose of class assignments.

Like most things, learning is a process and a skill that can be developed. For cognitive learning, the process involves two steps. *Step one is receiving* new

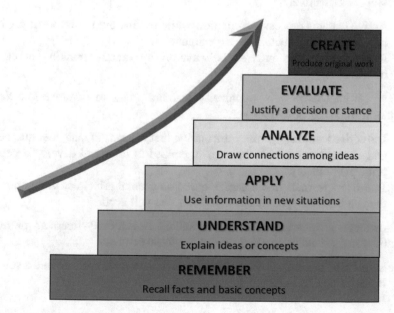

Fig. 2.2 Levels of intellectual skill in Bloom's Taxonomy typically used in engineering courses

information. *Step two is processing* the new information. Engineering education experts Rebecca Brent and Richard Felder [5] have shown that learners have preferences when receiving knowledge and processing knowledge.

Step One: When receiving knowledge, learners have preferences for (1) the *type of knowledge* they receive and (2) the *sensory input mode* in how they receive the information.

1. Type of knowledge—learners have preference between **sensing** and **intuitive** knowledge.

   • **Sensing** learners prefer facts, data, and real-world relevance. They like things that can be seen, heard, or touched.
   • **Intuitive** learners prefer abstract ideas, dreams, possibilities, and theories. They dislike repetitive things and look for meaning and variety.

2. Sensory input mode—learners have a preference between receiving **visual** or **verbal** knowledge.

   • **Visual** learners learn better through diagrams, pictures, graphs, flowcharts, sketches, videos, and demonstrations.
   • **Verbal** Learners learn better through spoken and written words.

Step Two: There are three aspects to processing new knowledge; (1) how the knowledge gets processed, (2) how the knowledge gets organized, and (3) how the learner progresses with the knowledge until understanding is achieved.

1. How the knowledge gets processed—learners prefer to process knowledge either **actively** or **reflectively**.

   • **Active** learners process information while writing out the information, talking about the information, or experimenting.
   • **Reflective** learners process information by quietly thinking about the information.

2. How the knowledge gets organized—learners prefer to organize knowledge either **inductively** or **deductively**.

   • **Inductive** learners organize information first in terms of data, measurements, and observations, which then get organized to infer and develop a general principle or theory.
   • **Deductive** learners first organize based on general principles or theories and then deduces the specific consequences that will occur.

3. Knowledge progression until understanding is achieved—learners prefer to achieve understanding in a **sequential** or **global** context.

   • **Sequential** learners prefer to follow a logical stepwise path to find a solution or understand knowledge.

- **Global** learners prefer to process knowledge at a high level and fill in the details as they make the connections. They tend to achieve understanding when all the seemingly random information "clicks" together.

This next activity helps us explore learning styles. However, it is important to remember that *all learners benefit from being exposed to all the learning styles*. Being in classes that are compatible with your learning preferences can help you easily learn. But being in classes that are incompatible with your learning preferences is also valuable because it will help you increase your capability to learn in that style. Therefore, expanding your skillset.

2.2.1. **Give an example of an intellectual activity for every level of Bloom's Taxonomy.**

2.2.2. **Complete the *Index of Learning Styles Questionnaire* found at the website https://www.webtools.ncsu.edu/learningstyles/. Write down your number score for each of the four dimensions of learning.**

2.2.3. **Based on the scores in the previous question and the website, what do the results mean to you? How do you plan to utilize the results when approaching your college classes?**

2.3 Improving the Learning Process

Since learning is a process, it can continually be improved. Metacognition is defined as the awareness and understanding of one's own thought processes [6]. To increase the metacognition of your learning follow the common and repeatable process shown in Fig. 2.3 [7]. When *planning* your learning make sure you answer questions such as why you are performing a task or learning a concept and make sure

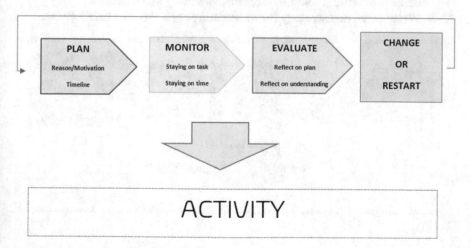

Fig. 2.3 Process to improve one's learning

you have a schedule, so you know roughly how much time it takes to complete a task. When *monitoring* your learning make sure you are staying on time with your work and reflect on your understanding. If understanding is low, seek help. When *evaluating* your learning, reflect on whether you planned work and you received the result you desired. If not, think about what can be done differently next time. Continually working through this repetitive process will build up your learning skills over time.

To be successful in earning an undergraduate engineering degree, it is critical that you avoid common mistakes students make early on in their schooling. If you have strategies to avoid the list given below, you will increase your chance of successfully completing the degree. This next sct of questions helps us to create approaches to improving our current learning strategies.

Mistake 1: Assume studying engineering will be just like high school
Mistake 2: Commit to too many extracurricular activities that grades begin to suffer
Mistake 3: Stay in house or dorm so much that little to no time is physically spent on campus
Mistake 4: Neglecting to study until the night before something is due
Mistake 5: Studying a topic only once during the semester
Mistake 6: Studying completely alone
Mistake 7: Come to lecture unprepared
Mistake 8: Avoid talking to professors
Mistake 9: Miss attending the majority of lectures
Mistake 10: Do not take notes during class or while watching at home video lectures
Mistake 11: Fail to complete homework assigned
Mistake 12: Never ask anyone on campus or at home for help

2.3.1. **Pick one learning activity you have encountered so far in college (homework, test, project, etc.) and fill in the blank Fig. 2.3. Make sure you include planning, monitoring, and evaluating mechanisms.**

2.3.2. **Write down one strategy for overcoming each mistake listed above.**

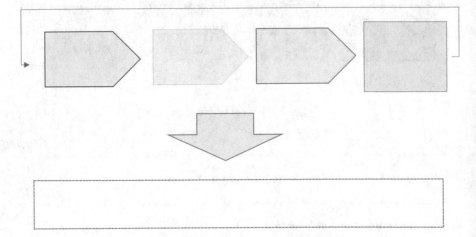

End of Chapter Problems

IBL Questions

IBL1: Choose one class you are currently taking and answer the following questions.

1. What is the style of teaching in the class?
2. Who is responsible for the learning and what technology facilitates the learning?
3. What is the method of teaching in the class?
4. What is the type of knowledge being taught?
5. What mode is the knowledge presented in?
6. How is the presentation of knowledge organized?
7. How do students participate?
8. What viewpoint does the knowledge come from?
9. Explain whether you feel the style of teaching, method, and instruction elements facilitate your learning?
10. Discuss three steps you can take to adjust to the teaching style to improve your learning.

IBL2: After completing the activities in Section 2.2, go to the link, read the document, and answer the following questions:https://www.engr.ncsu.edu/wp-content/uploads/drive/1WPAfj3j5o5OuJMiHorJ-lv6fON1C8kCN/styles.pdf

1. Are you a more reflective or active learner?
2. Write down three strategies that the website gives for how you can help yourself based on your learning style.
3. Are you a more sensing or intuitive learner?
4. Write down three strategies that the website gives for how you can help yourself based on your learning style.
5. Are you a more visual or intuitive verbal?
6. Write down three strategies that the website gives for how you can help yourself based on your learning style.
7. Are you a more sequential or global learner?
8. Write down three strategies that the website gives for how you can help yourself based on your learning style.

Practice Problems

1. From section 2.1, rank the six teaching methods listed in order from your most preferable to least preferable. Explain the reason for your rankings and for the bottom two ranked teaching methods, name three strategies you can employ to learn from those even though you do not prefer them.

2. Write an essay about a previous class where you had a great learning experience. Describe the style of teaching, method, and instruction elements for that class. Explain what you learned and why you liked the class.
3. Write an essay about a previous class where you had a poor learning experience. Describe the style of teaching, method, and instruction elements for that class. Explain what you learned and why you did not like the class. Reflect on whether there were behaviors or strategies you could have changed to have had a better experience.

References

1. Teaching Methods, teach.com. viewed May 18, 2021.
2. Felder, Richard M. and Silverman, Linda K., (1988) "Learning and teaching Styles in Engineering Education," Engineering Education, v. 78(7), pg. 674-681.
3. Bloom, B. S.; Engelhart, M. D.; Furst, E. J.; Hill, W. H.; Krathwohl, D. R. (1956). Taxonomy of educational objectives: The classification of educational goals. Handbook I: Cognitive domain. New York: David McKay Company.
4. Anderson, L.W. and Krathwohl, D.R., (2001) A Taxonomy for Learning, Teaching, and Assessing: A Revision of Bloom's Taxonomy of Educational Objectives, Complete Edition, Longman, New York, NY.
5. Felder & Brent, (2016), Teaching and Learning STEM: A Practical Guide, Jossey-Bass, San Francisco, CA. ISBN-13: 978-1118925812.
6. Oxford English Dictionary. Oed.com.
7. Landis, R.B., (2007), Studying Engineering: A Road Map to a Rewarding Career, Discovery Press, Los Angeles, CA.

Chapter 3
Essential Skills for Learning

Abstract Establishing yourself as an effective engineer begins in the classroom. Throughout your time in college, you will encounter different subject matter taught by different faculty, and one thing is certain, all professors teach in different ways. One professor's teaching style (recall the discussion in the previous chapter) might not align with your learning style (also discussed in Chap. 2). However, you cannot let this deter you from succeeding in learning the course material. Having courses taught in different ways is an opportunity for you to become agile in your aptitude to learn. This chapter introduces many longstanding strategies to apply that will improve your ability to get the most out of your learning, regardless of the teaching style in the class.

By the end of this chapter, student will learn to:

- Develop strategies for effective course and lecture preparation.
- Develop strategies and skills for effective learning during class lecture.
- Develop skills for reading for comprehension of technical material.
- Develop skills to prepare for, and successfully take exams.
- Develop strategies to effectively organize and manage your time.
- Understand the benefits professors can provide, and explain strategies for developing effective relationships with professors both inside and out of the classroom.
- Understand the benefits of utilizing your peers effectively.
- Understanding the benefits of Institutional Resources (career services, tutoring, etc.)

3.1 Semester and Course Preparation

Course preparation is associated with the beginning of a semester and has been equated to the start of a marathon shown in Fig. 3.1. The goal should be that when the race begins (meaning the first day of classes); you want to be off and running at your best pace. Much like training prior to running marathon, to start the semester

© Springer Nature Switzerland AG 2022
M. Blum, *An Inquiry-Based Introduction to Engineering*,
https://doi.org/10.1007/978-3-030-91471-4_3

Fig. 3.1 Course preparation has been equated to the start of a race

strong, mental and physical preparation is required. This guided inquiry helps to discern goals and develop plans to start your semester and new courses in the best way possible.

3.1.1 **How do you know that you are in the appropriate classes at the beginning of a semester?**

3.1.2 **What items are needed for the first day of a class?**

3.1.3 **Based on the previous questions and your experience, come up with goals you want to set for yourself so that you are prepared for the beginning of each semester.**

3.1.4 **Before the start of a semester, three typical goals are (1)** *Be in all appropriate courses* **(2)** *Have textbooks and materials, and* **(3)** *Be mentally prepared***. For each goal, develop at least one strategy you can employ to accomplish that goal.**

3.1.5 **For goal (2) in the previous question, list items that need to be bought and propose a timeline for purchasing.**

3.1.6 **If you do not know what specific items are needed for a class, who should you ask?**

3.1.7 **Explain some techniques you use to mentally prepare yourself for the semester. (For example, address things like (1) being distracted at the start (2) being complacent because the semester starts off slowly (3) being overwhelmed by a difficult schedule)**

3.2 Lecture Preparation

Lecture preparation is a tactic that is underutilized by students. It takes a relatively small effort but can have a large benefit. It is like warming up before you work out or play a sports game. If you take a little time to prepare before lecture, you will enter the lecture with more interest and focus, and thus leave having learned significantly more than, if you went in unprepared.

One of the most useful items to prepare for class is the course syllabus. The course syllabus is a document provided by the instructor of a course. It communicates specifics about the course, defines expectations for the class, and the responsibilities each student has to that class. The syllabus is a treasure of information. It is essentially a contract between you and the professor about expectations for the class each semester. It is not just what the professors expect from *you* students, but also what *you can expect from the professor*. Information in the syllabus includes many important items such as how your grade is calculated, exam dates, and a course calendar. An example of a syllabus can be found in Fig. 3.2. Remember to study the syllabus thoroughly. Keep it all semester and before you ask the professor a question, read it! This guided inquiry has us identify strategies to help prepare prior to each lecture.

3.2.1 **Look at the syllabus in Fig. 3.2. What is the exam contention policy for that Course? Explain what an exam contention policy is.**

3.2.2 **Read the syllabus for your specific class. What is the most important piece of information on the syllabus? Explain why you think it is important.**

3.2.3 **Read the syllabus for your specific class. What is the most important Policy on the syllabus? Explain why you think it is important.**

3.2.4 **Write down as many strategies as you can think of that you can employ prior to class (maybe the night before) that will mentally prepare you for class.**

For the strategies you discussed, try at least one of these for at least the first 2 weeks of classes.

3.3 Learning During Lecture

Once you have warmed up and prepared for lecture, consider lecture game time! Lecture is not a passive experience, where students inactively sit and magically absorb the course material. Students need to take an active role in their learning process and thus lecture. Experts suggest [1] several best practices to employ in order to get the most out of lecture. They include:

- Sit near the front of the class.
- Work to concentrate on the material being presented.

Syracuse University   **College of Engineering and Computer Science**
ECS 101   **Fall 2018**

Course:	Introduction to Engineering and Computer Science (Mechanical Engineering)
Lecture:	Tuesday & Thursday 9:30am – 10:50am Newhome Bld. Room 141
Laboratory:	(M017) Wednesday 10:35am – 11:30am Blink Hall Room 143 (M024) Wednesday 11:40am – 12:35pm Blink Hall Room 143 (M025) Wednesday 12:45pm – 1:40pm Blink Hall Room 143
Instructor:	Dr. Amy Bloom Office Number: 200 Blink Hall Phone: (315) 443-4018 e-mail: mbloom@syr.edu Office Hours: Mondays 10am – 12pm (or by appointment)
Teaching Assistants:	Todd Fitzherald Email: jtfitzhe@syr.edu Office hours: Thursdays 11am – 1pm Location: Blink 217A
Recommended Texts:	*Saeed Moaveni, Engineering Fundamentals: An Introduction to Engineering, Thompson (2005). (ISBN:0-534-42459-7)* *Raymond B. Landis, Studying Engineering: A Road Map to a Rewarding Career, Discovery Press (2007). (ISBN:978-0-9646969-2-1)*
Online Resources:	www.blackboard.com
Learning Outcomes:	As a first course in engineering materials for students, the primary objective of this course is to introduce students to the engineering profession and the skills necessary to be successful in the pursuit of and engineering degree. Specifically, at the completion of this course the student should:

- Develop an ability to identify, formulate, and solve complex engineering problems by applying principles of engineering, science, and mathematics [ABET Outcome 1].
- Develop an ability to apply engineering design to produce solutions that meet specified needs with consideration of public health, safety, and welfare, as well as global, cultural, social, environmental, and economic factors [ABET Outcome 2].
- Develop an ability to communicate effectively with a range of audiences [ABET Outcome 3].
- Develop an ability to recognize the ethical and professional responsibilities in engineering situations and make informed judgements, which must consider the

Fig. 3.2 Example Syllabus from an Introduction to Engineering Class

impact of engineering solutions in global, economic, environmental, and societal contexts [ABET Outcome 4]
- Develop an ability to acquire and apply new knowledge as needed, using appropriate learning strategies [ABET Outcome 7]
- Become familiar with the various fields of Engineering and Computer Science and become engaged in the SU E&CS Community
- Begin their Career preparation, learn strategies and college resources for success as a college student

Grading:

Homework	10%
SEM100	10%
Manufacturing Design Project	15%
Units Quizzes	20%
Final Exam	25%
Open Design Lab Book	10%
Open Design Project Performance	10%

Grading continued:

The course grades will be determined nominally as follows:

≥ 93	93 – 80	79 – 74	73 – 70	69 – 66	65 – 60	59 – 50	< 50
A	A- or B+	B	B- or C+	C	C-	D	F

- If you fall into one of the "gray areas," your grade will be determined by whether your performance has improved or remain consistent (higher grade) or gotten worse (lower grade), depends particularly on quizzes and the final exam.
- The grey area cutoffs are determined at the discretion of the instructor.

Homework Assignments:

- Late homework will **NOT** be accepted except in the case of professional/SU approved athletic travel, or appropriately documented serious emergency. Contact the instructor *before the due date* to arrange an alternative accommodations.

Exams Dates:

- Unit Quiz #1 Scheduled for: **Tuesday, September 18, 2018**
- Unit Quiz #2 Scheduled for: **Thursday October 18, 2018**
- Unit Quiz #3 Scheduled for: **Thursday November 15, 2018**
- Final Exam Scheduled for: **Monday December 10, 2018 – 12:45pm – 2:45pm**

Exam Contention Policy:

- There is no time limit on problem explanation.
- Dr. Bloom is open to grade discussion approximately *1 week* after grade is returned.
- In order to discuss/contend a grade, students must send Dr. Bloom an email expressing grievance and desire to meet, then see Dr. Bloom at office hours or set up an appointment via email.

Academic Integrity:

- Discussion of homework problems is encouraged; however each student is expected to submit his or her own, independent solutions. Working together, asking questions of classmates, or assisting others on quizzes or exams is prohibited.
- Syracuse University's Academic Integrity Policy holds students accountable for the integrity of the work they submit. Students should be familiar with the policy and know that it is their responsibility to learn about course-specific expectations, as well as about university policy. The university policy governs appropriate citation and use of sources, the integrity of work submitted in exams and assignments, and the veracity of signatures on attendance sheets and other verification of

Fig. 3.2 (continued)

participation in class activities. The policy also prohibits students from submitting the same written work in more than one class without receiving written authorization in advance from both instructors. The presumptive penalty for a first offense by an undergraduate student is course failure, accompanied by a transcript notation indicating that the failure resulted from a violation of Academic Integrity Policy. The standard sanction for a first offense by a graduate student is suspension or expulsion. For more information and the complete policy, see http://academicintegrity.syr.edu/academic-integrity-policy/

Disability-Related Accommodations:
- If you believe that you need accommodations for a disability, please contact the Office of Disability Services (ODS), http://disabilityservices.syr.edu, located in Room 309 of 804 University Avenue, or call (315) 443-4498, TDD: (315) 443-1371 for an appointment to discuss your needs and the process for requesting accommodations. ODS is responsible for coordinating disability-related accommodations and will issue students with documented Disabilities Accommodation Authorization Letters, as appropriate. Since accommodations may require early planning and generally are not provided retroactively, please contact ODS as soon as possible.

Religious Observances Policy:
- SU religious observances policy, found at http://supolicies.syr.edu/emp_ben/religious_observance.htm, recognizes the diversity of faiths represented among the campus community and protects the rights of students, faculty, and staff to observe religious holidays according to their tradition. Under the policy, students are provided an opportunity to make up any examination, study, or work requirements that may be missed due to are religious observance provided they notify their instructors before the end of the second week of classes. For fall and spring semesters, an online notification process is available through MySlice-StudentServices-Enrollment-MyReligiousObservances from the first day of class until the end of the second week of class.

Diversity Policy:
- It is my intension that students from all backgrounds and perspectives will be well served by this course, and that the diversity that students bring to this class will be viewed as an asset. I welcome individuals of all ages, backgrounds, beliefs, ethnicities, genders, gender identities, gender expressions, national origins, religious affiliations, sexual orientations, socioeconomic background, family education level, ability – and other visible and non-visible differences. All members of this class are expected to contribute to a respectful, welcoming and inclusive environment for every other member of the class. Your suggestions are encouraged and appreciated

Fall 2018 Schedule**

Week	Dates	Tuesday	Thursday
1	8/28 & 8/30	Introduction to Engineering	Study Skills I
2	9/4 & 9/6	History of Engineering at Syracuse University	Study Skills II
3	9/11 & 9/13	Homework Skills	Scientific Units I
4	9/18 & 9/20	Unit Quiz# on 9/18	Units II & Measurements
5	9/25 & 9/27	Time	Length
6	10/2 & 10/4	Mass	Career Services Guide
7	10/9 & 10/11	Problem Solving I	Problem Solving II
8	10/16 & 10/18	Engineering Ethics	Unit Quiz#2 on 10/18
9	10/23 & 10/25	Introduction to Nanoscience	Introduction to Nanoscience
10	10/30 & 11/1	Force I	Force II
11	11/6 & 11/8	Current	Energy
12	11/13 & 11/15	Communication Skills	Unit Quiz#3 on 11/15
13	11/20 & 11/22	Thanksgiving Break	
14	11/27 & 11/29	Math I	Math II
15	12/4 & 12/6	Math III	Final Design Project Demonstration*

*Final Design Project Demonstration Day is Thursday December 6, 2018, from 9:30am – 10:50am in Hannigan Gymnasium.
**All policies, dates etc. given are subject to change by class announcements.

Fig. 3.2 (continued)

Table 3.1 A list of Poor Listening Behavior and Good Listening Behavior

Poor Listening Behavior	Good Listening Behavior
Tunes out quickly if they think the topic is uninteresting and boring.	
Tunes out if the delivery of material is not how they like it.	
Listens for facts and details.	
Brings little energy to the listening process.	
Reacts with opposing views to new ideas. Listens to *themselves* when they hear something they do not agree with.	
Easily bothered by distractions.	
Prefers light leisure material; resists difficult material.	
Interrupted by emotionally charged words or ideas.	
Daydreams with slow speakers or gaps in presentation.	

- Practice *active* listening behavior (see Table 3.1).
- Take thorough notes. An example of a suggested note-taking system is shown in Fig. 3.3.
- Do not be afraid to ask many questions.
- Limit electronic devices unless necessary for class.

The next guided inquiry helps us to explore these tactics in more detail.

3.3.1 **Explain why sitting near the front of the classroom will increase a student's learning during lecture.**

3.3.2 **List some strategies to employ to help stay focused and concentrate during lecture.**

3.3.3 **Table 3.1 lists a table of Poor Listening Behavior. Fill in the Good Listening Behavior side of the table that reverses the poor behavior.**

3.3.4 **In Table 3.1, list any item where you would describe yourself as showing Poor Listening Behavior.**

3.3.5 **For each item listed, describe a strategy you will use to change your habit.**

3.3.6 **For Note-taking, list the pros and cons of using a spiral notebook.**

3.3.7 **For Note-taking, list the pros and cons of using a three-ring binder.**

3.3.8 **For Note-taking, list the advantages and disadvantages of taking notes on a computer.**

3.3.9 **When asking questions in class, is it okay to ask a professor "I am confused about the point you just made, could you explain it again?"**

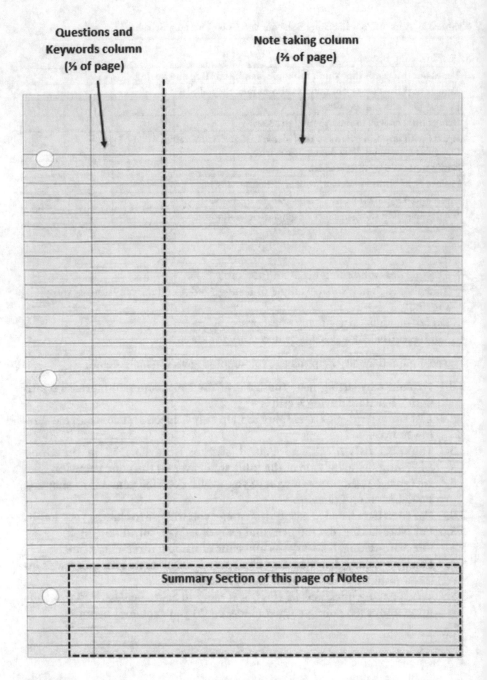

Fig. 3.3 A widely utilized strategy for note-taking is the Cornell Note-Taking System [2]. It was created by an education professor at Cornell University, Walter Pauk, in the 1940s. It is a systematic format for condensing and organizing notes. The method suggests that students divide their paper into two columns: the note-taking column (usually on the right) is twice the size of the questions/keyword column (on the left). Finally, at the bottom of each page five to seven lines should be left for students to summarize that section of notes

3.4   Reading to Learn

While pursuing an engineering degree involves design and hands-on work, much of the learning that will occur requires you to read, both inside and outside the classroom. Class notes, textbooks, and journal articles are some examples of the highly technical nature of the reading that you will encounter. Reading technical literature is unlike reading fictional literature for entertainment because leisurely reading can typically occur at a relatively fast pace. While understanding math, science, and engineering content take a slow, repetitive process where the reader needs to be highly engaged.

Consider reading a technical passage in three main stages, and at each of those stages there are steps you can take so that the reading you engage in is *active*, which better helps you comprehend the material. Below describes each stage and step.

Stage 1: Before You Read

> Step 1: Establish a purpose for reading. Either mentally acknowledge it, or write it down.
> Step 2: Skim the reading. Note how many pages it is. Check the headings and subheadings in each section. Then read the introductory paragraph.
> Step 3: Make a list of questions to be answered by the reading. You can use the main headings and subheadings to help develop questions.

Stage 2: While You Read

> Step 1: Always use a pencil and notepaper. While you are reading, actively summarize concepts, sketch graphs, perform derivations, and write down questions about anything you do not understand.
> Step 2: Take your time to read slowly, focusing on concepts, and not memorizing examples.
> Step 3: Periodically, stop and paraphrase what you have read using your own words. Do not just recite exactly from the text.

Stage 3: After You Read

> Step 1: Report the answers to the questions you prepared before you started reading in your own words. If you cannot answer the questions, reread where needed.
> Step 2: Within 1 day of reading, review the material; recite the main points and answer pertinent questions again. Continue to review throughout the semester.
> Step 3: Solve problems that are based on the reading. Always start with assigned problems (typically homework), but then work on other example problems to reinforce the concepts.

The next guided inquiry helps us to understand the importance of this procedure.

> *"There once was a little boy named Carl Gauss. He had a very lazy teacher who did not want to teach one morning, so the teacher gave the class an assignment to add the numbers 1-100. To his surprise, Carl came up with the answer (5,050) in about a minute. The teacher thought Carl had cheated and asked him to explain how he had come up with his answer so quickly.*
>
> *Carl noticed very quickly that the sum was the same when he added the first and last number, the second and second-to-last number, the third and third-to-last number, and so on. He figured out that because there were one hundred numbers, there would be fifty pairs that were equal to one hundred and one. The sum of the numbers 1-100 would be equal to the number of pairs multiplied by the sum of each pair. Karl was able to use what he knew about numbers to solve what seemed like a complicated assignment in a short amount of time."*

Fig. 3.4 Short story about German mathematician and physicist Carl Gauss [3]

3.4.1 **Explain why focusing on concepts, rather than memorizing examples will increase comprehension of the technical material.**

3.4.2 **Explain why it is important to paraphrase the reading, rather than reciting exactly from the text.**

3.4.3 **Write out your timeline for how often you plan to review material (tip: plan backward from any major milestones such as an exam or project).**

3.4.4 **How many example problems per day should be practiced?**

3.4.5 **Read the passage shown in Fig. 3.4. Apply the reading methods you just learned to the passage, then explain the theorem in your own words.**

3.4.6 **Referring to the passage in Fig. 3.4, write the theorem in the language of mathematics (namely, make it an equation).**

3.4.7 **Using the answers to questions 3.4.5 and 3.4.6, add all consecutive numbers from 20 to 27.**

The exercise you just went through with Carl Gauss's method for adding consecutive numbers is an example of technical reading and analysis. After a reading such as this, finally, remember to take time to reflect on the reading. Think about the level of understanding of the concept you achieved. Reflect on if you are curious about any topic beyond the reading (for example, who is Carl Gauss and what else did he do?). Throughout your academic career, most reading in your math, science, and engineering classes your encounter will be of this nature.

3.5 Preparing for Exams

Another inventible part of studying to become an engineer is exams. Written tests are one of the prime ways Professors evaluate student learning. There is a saying that *"studying hard is the best form of luck there is."* This is very true. It is always sad as an instructor when we hear students talk about how they stayed up all night studying for an exam. This student is someone who did not properly manage his or her time. If you are consistently working on your course material then when you study for a test, you will be reviewing material (instead of learning it for the first time). This inquiry helps us appreciate preparing well for exams and examines strategies that will help students prepare for exams.

3.5.1 **Describe the main difference between test taking and all other forms of learning in the class.**

3.5.2 **List at least three *study tactics* you use to prepare for an exam.**

3.5.3 **For each tactic listed in question 3.5.2, develop a timeline for the activity (the week(s) before the exam, the night before the exam, etc.)**

3.5.4 **When using the tactics listed in the above question, list at least three *sources of information* from the course that you study (example, textbook).**

3.5.5 **List at least three rituals (*not studying*) you have to prepare the night before and/or the day of an exam.**

3.6 Successfully Taking Exams

Once you have effectively prepared for the exam, it is important to develop a process for taking the exam. Having your own process to follow will help you to remain calm and focused during an exam. Given here are a few test-taking strategies:

1. When you start, size up the test.

   - Read directions carefully.
   - How many sections is it?
   - What type of sections/questions does it have?
   - How much is each section worth?

2. Answer the easy problems first.
3. Complete a problem before leaving it.
4. Try to answer all of the questions, even if you are unsure of your answer.
5. Be aware of the time.
6. Ask Professor questions if anything is unclear.
7. Use all of the time given, so never leave early!

The next guided inquiry helps us to examine some of these strategies more in detail.

3.6.1 **For Strategy 1, list the types of questions you might see on an exam.**

3.6.2 **For Strategy 1, explain why it is important to read the directions carefully before you begin the exam.**

3.6.3 **For Strategy 2, explain why you should answer the easy problems first.**

3.6.4 **Explain the importance of Strategies 3, 4, 5, 6, and 7.**

3.7   Time Organization and Management

Time is the one resource that everyone has in equal amounts. Everyone has the same amount of hours in a day, days in a week, weeks in a month, and months in a year. One of the things new college students always mention about the transition from high school to college is the difficulty of self-regulating their responsibilities and managing their time. In college, no one is there to wake you up in the morning, take you to school, or do your laundry. You have to rely on yourself to complete all of your responsibilities and ultimately be successful. This may seem overwhelming at first, but with a strong time management system in place, you can be a person who accomplishes a great amount and still have time to enjoy all of the frivolities of the college experience. Before a system can be developed, first we need a feel for the amount of time we have. This guided inquiry helps us do just that.

3.7.1 **How many hours are in 1 week?**

3.7.2 **Make a list of weekly activities that college students perform.**

3.7.3 **If to get enough rest to stay healthy you needed 7 h of sleep per night, how many hours should you be sleeping each week?**

3.7.4 **If you spend 1 h for each meal per day, and you eat three meals per day, how many hours should you spend eating per week?**

3.7.5 **If you spend 1 h each day showering, and then roughly 3 h over the weekend doing laundry and cleaning, how many hours should you spend on personal grooming/hygiene per week?**

3.7.6 **A typical full time engineering student takes nominally 16 credits per semester. If each credit equates to an hour of class time each week. How much time is spent in the classroom each week?**

3.7.7 **It is recommended that students spend 3 h working outside of the classroom per class credit (this includes homework, reading, watching videos, online material, studying, etc.). Based on this, how much time should you spend working outside of the classroom on course items?**

3.7.8 **Based on Questions 3.7.6 and 3.7.7, how many hours per week should you be spending working on class items?**

3.7.9 **Based on Questions 3.7.1 thru 3.7.8, how many hours per week should you have as "free" time?**

Now that we appreciate how we have to spend our time, it is important to develop a plan for organizing where and when you complete these activities. When working on time management, remember the phrase "planning is no PRANK." This will help you remember the key steps. Let us look at them individually.

   P = Priority Management

In 1989 a self-help book was written called The 7 habits of Highly Effective People [4]. The third highly effective habit deals with priority management. Highly effective people organize their time and execute their activities based on groups of what is *important* and what is *urgent*. Important items are based on personal values, and urgent items are things that require immediate attention. A matrix can be created to group action items based on four quadrants shown in Fig. 3.5:

Fig. 3.5 Matrix of Importance versus Urgency to set up when deciding how to manage your time

	URGENT	NOT URGENT
IMPORTANT	Quadrant I *urgent and important* **DO**	Quadrant II *not urgent but important* **PLAN**
NOT IMPORTANT	Quadrant III *urgent but not important* **DELEGATE**	Quadrant IV *not urgent and not important* **ELIMINATE**

Q1: Urgent and Important—important deadlines and crises. These items you complete right away!

Q2: Not urgent but important—long-term development. These items you can set a plan to complete.

Q3: Urgent but not important—These items are sometimes known as "distractions with deadlines."

Q4: Not urgent and not important—These are often frivolous items that can distract from one's ultimate goals.

The order is important; after completing items in Q1, we should spend the majority of our time on Q2, and try to avoid Q3 and Q4. However, many students fall into the trap of spending too much time in Q3 and Q4.

R = Recreation and Errands—Remember to make time for fun chores

We discovered in the last guided inquiry that even though college is busy, there is ample "free" time. Although college is a time to learn your future trade, it is also a time to develop life skills and have experiences. Your time management has to account for responsibilities such as laundry, cleaning, and paying bills. Also, remember to take care of yourself physically; get enough sleep and schedule physical activity (gym time, sports, taking a walk, etc.). Your mental capability is greatly affected by your physical status. Finally, life experiences are good for students. Schedule some time to engage in the college community. Depending on your interests (art, music, theater, athletics, volunteering, fraternity, sorority, or professional societies) there are opportunities for you to engage at college.

A = Anticipate delays

Murphy's Law is a saying that goes "Anything that can go wrong will go wrong." Remember this when planning your time. Make sure to account extra time for setbacks such as running out of paper, pens or pencils, computers breaking, or getting sick. Also, make sure to budget time for the "finishing touches" of work. Remember to double-check all of your work, proofread papers, and polish projects. You can also enlist the help of friends or professors with the final changes.

N = No—It is okay to turn something or someone down

In order to stay focused on your goals, you have to become comfortable with saying "no." When someone requests something from you, or you are interested in some activity ask yourself "Will this help me reach my goal?" and "If I say yes, where will I fit it in my plan?" Remember that you are in college to get an education and become an engineer. Make an effort to schedule academics first, then social activities. Work to say "no" if you are not finished with your week's coursework. This will also help you stay out of the Q3 area when setting a plan to manage your time.

K = Keep it simple

There is a saying "Rome wasn't built in a day." The same concept applies to your goals and managing your time. Understand that some goals are large and will take time to accomplish. These long-long-term goals can be broken down into smaller short-term goals. Stating your goals simply and directly will help you focus your priorities, and thus plan your actions to accomplish your goals.

Now that we have learned about time managing strategies, this guided inquiry helps us reflect on our individual planning process.

3.7.10 **List items that you will have to accomplish each week.**

3.7.11 **Based on the answer to question 3.7.10, make importance versus urgency chart (like the one shown in Fig. 3.5) for each of the items you listed.**

3.7.12 **Name one long-term goal you want to accomplish this semester.**

3.7.13 **Make a list of short-term goals that will need to be completed in order to accomplish the goal set in question 3.7.12.**

3.7.14 **List three benefits of scheduling your study time.**

3.7.15 **List all the things you consider when deciding how much time to budget for studying for a class.**

3.7.16 **List how many hours you plan to study for each course.**

3.7.17 **Describe how you plan to organize those hours during each class week. (Think about where you are going to keep your schedule.)**

3.8   The Student/Professor Relationship

Most students view college professors as unapproachable people and students only want to interact with them during the specific class they are teaching the student. However, a professor's role to contribute to your educational experience is not just limited to classes. Understanding how to make effective use of your professors can greatly improve a student's college experience and future prospects. In order to understand how to interact with your professors, it is important for you to have a good understanding of what professors actually do. By doing so, hopefully, you will be sensitive to the demands placed on them. This will help you build more effective relationships with them. First off, professors do more than teach. They are expected to perform in three primary categories:

Teaching

- Activities include classroom teaching, curriculum development, laboratory development, academic advising, supervising student projects, etc.

Research

- Accomplishments include creating and organizing new knowledge: publications, journals, textbooks, software, presentations, participation in professional societies, and funds to support all of this!

Service

- Activities include community and campus involvement, faculty governance (department, college, and university committees), consulting, etc.

Although professors are very busy, one of their primary roles is student support. That support can come in many forms including the following list:

- One-on-one instruction (Office Hours)
- Academic advising, career guidance, personal advice
- Monitor your progress; hold you accountable
- Give you the benefit of the doubt on borderline grades
- Help you find a summer job
- Hire you on their research grant
- Serve as a reference
- Nominate you for scholarships or academic awards

As you can see there are so many benefits to what we can offer you, but each student has to take responsibility to get to know their professors. Professors cannot help you if they do not know you.

Now that you understand the importance of developing relationships with your professors, let us discuss strategies of how to develop a relationship with them. First, you must take the initiative to establish the relationship. A list is suggested of *Behaviors to Display* both in and out of class and *Behaviors to Avoid* are shown in Table 3.2. By utilizing these strategies you will increase your chances of your professors learning your name and becoming a supportive resource on campus for you. This inquiry helps us to reflect and develop a strategy to develop relationships with college professors.

3.8.1 **Look at the bulleted list given above. List your top three ways you want support from your professors and explain why those are important to you.**

3.8.2 **Based on your list in 3.8.1, explain what actions you are willing to take to ensure that your professors will support you for your list.**

3.8.3 **Looking at Table 3.2, list your top three *Behaviors to Avoid* that you have caught yourself doing.**

3.8.4 **Based on your list in 3.8.3, explain what actions you will take to change your habits.**

Table 3.2 A list of behaviors to display and to avoid when interacting with college professors

Behaviors to display	Behaviors to avoid
Sit in the front of the class	Sleeping in and being late to class
Pay attention in class	Talking during class while the professor is talking
Visit office hours often	Trying to complete homework during class
Ask questions in class	Using cell phones/tablets/computers in class (if not told otherwise)
Be on time for class	Leaving class early
Address them appropriately (Professor or Doctor)	Failing to do the assignment
Be kind and patient	Correcting Professor's mistakes in an antagonistic tone
Relax when speaking	Complaining assignments and exams are too hard
	Complaining grade is unfair
	Taking disagreement personally

3.8.5 **In Table 3.2, there are two blank spaces under *Behaviors to Display*. Write in two of your own behaviors you think you should display when interacting with your professors.**

The final important thing about communication with your professors is email. Outside of class time, email is probably the most powerful tool for communicating with your professors. Used responsibly, email is a venue for introducing yourself, setting up a meeting, asking a question, or expressing a concern. However, there is a high potential for students to misuse email, or miss an opportunity because of using email unprofessionally. The next inquiry helps us to think about the proper protocols for emailing professors.

3.8.6 **When emailing a professor, why is it important to write the email from your college email account?**

3.8.7 **When emailing a professor, why is it important to include a description (such as the course name or number) in the email subject line?**

3.8.8 **When emailing a professor, what is an inappropriate greeting?**

3.8.9 **When emailing a professor, what is an appropriate greeting?**

3.8.10 **Make a list of things to avoid when emailing your professor.**

3.8.11 **Make a list of things to always remember to do when emailing your professor.**

3.8.12 **Should you text your professor or reach out to them on social media? Explain why or why not.**

3.9 Working with Your Peers and in Teams

The ability to work with in a team and with peers is one of the most critical skills an engineer can develop. No matter how talented an engineer you are, you will never have all of the answers. You need your peers and you can always learn something from others. During your studies, and ultimately in the workplace, you will have to complete assignments individually, but many other assignments will be accomplished by working in an informal group, or an organized team. In this inquiry, we explore the difference between working in a group and working in a team.

3.9.1 **In your own words, define what is meant by a *group*.**
3.9.2 **In your own words, define what is meant by a *team*.**
3.9.3 **Based on your answers to questions 3.9.1 and 3.9.2, explain the similarities between a team and a group.**
3.9.4 **Based on your answers to questions 3.9.1 and 3.9.2, explain the main differences between a team and a group.**

Now that we understand the difference between working in a group and in a team, the next set of questions discusses some strategies to work efficiently in a group.

3.9.5 **Do you spend some fraction of your study time on a regular basis studying with at least one other student? Or do you spend 100% of your time studying alone?**
3.9.6 **If you study alone, explain why you do not study with other students (what do you think is hindering you from studying with other students). If you study with other students reflect on how it is working for you.**
3.9.7 **List what you think are the benefits of studying in a group.**
3.9.8 **When studying in a group, what percentage of a student's overall studying should be done in a group?**
3.9.9 **When studying in a group, what is the ideal size of a study group?**
3.9.10 **When studying in a group, what can be done to keep the group from getting off task?**

When you work in a study group, or work independently, you are free to choose when and how you work. When you are part of a team working to accomplish the same goals, there is the potential for great productivity and diversity of idea, but there is also a high potential for conflict. Conflict can be both productive and unproductive. In order to mitigate unproductive conflict, the team needs to conduct themselves in the most professional ways possible. Listed below are some strategies to help manage teamwork. To get the team off to a strong start four things need to be established early;

Purpose: A successful team establishes a purpose. Each team needs to discuss their goals, outcomes and come to a consensus about acceptable and unaccepted individual and team behavior.
Process: A successful team establishes a process. Teams need to discuss and develop rules for decision-making, communicating with each other, roles and responsibilities, how the work will be distributed and expectations for participation of each team member.
Progress: Teams need to establish a means of measuring the team progress throughout the project.

Accountability: Teams need to establish a procedure for evaluating the performance of each team member and develop accountability measures. Namely, repercussions if deadlines and responsibilities are not met.

The next inquiry discusses some strategies to work efficiently in a team.

3.9.11 **When developing a team process, list some ways that teams can make decisions.**

3.9.12 **When developing a team process, list some ways that teams can communicate with each other.**

3.9.13 **When developing a team process, what are the roles and responsibilities that team members can hold during a project?**

3.9.14 **Explain what steps should be taken by a team before a team meeting?**

3.9.15 **Explain what steps should be taken by a team during a team meeting?**

3.9.16 **Explain what steps should be taken by a team after a team meeting?**

3.9.17 **If there is a conflict on a team, explain some strategies that members could use in order to resolve the conflict in a professional manner.**

3.9.18 **If a team member keeps missing deadlines for team assignments and not attending team meetings, explain a strategy for how the team would talk with this student.**

3.9.19 **Describe an approach to use when evaluating team members performance.**

3.10   Benefits of Institutional Resources

Every college institution offers numerous academic and personal services to support student's education. These services are paid for through your tuition and student fees. Like most other things in college, it is your responsibility to seek out these services. They are for your benefit, but you must take initiative to find them and use them when you need them. In general, you will have access to resources such as

- Tutoring Services
- Extra recitation or Problem-Solving Workshops
- Student Support Services
- Career Services
- Computer labs
- Academic Advising
- Registrar's Office
- Bursar's Office
- College Catalog
- Library

This guided inquiry lets us explore the services on campus.

3.10.1 **Explain what the Bursar's Office and Registrar's Office do.**

3.10.2 **Explain where your Bursar's Office and Registrar's Office are on campus.**

3.10.3 **Explain what the tutoring options are on campus.**

3.10.4 **List where the main computer clusters are that you will use on campus.**

3.10.5 **Explain what support Careers Services provides for students.**

3.10.6 **Where is the Career Services office on campus?**

3.10.7 **What is the function of the Student Support Service?**

3.10.8 **Explain where the Student Support Service offices are on campus.**

3.10.9 **Explain how you plan to interact with the Student Support Services on campus.**

End of Chapter Problems

IBL Questions

IBL1: Choose one professor whose class you currently are taking and attend their office hours. Write down their answers for the following questions.

1. Where did you attend college as an undergraduate student?
2. How did you become interested in this career path and why did you choose it?
3. What is your favorite part of your job?
4. What is your favorite class to teach?
5. What are your office hours?
6. What is your teaching style?
7. How can I perform well in your class?
8. If I begin struggling in your class, what resources are available to me?

IBL2: Find two other people in class, sit together, and ask them these questions.

1. Where are you from?
2. Why did you attend this school?
3. Why did you choose this major?
4. What are your extracurricular interests?
5. What other classes are you currently taking?
6. When do you have breaks in your schedule?
7. Can we find one similar break time to get together each week to study for this class?

Practice Problems

1. Construct an email to your professor for this class, briefly introduce yourself and ask one question about the class.
2. Construct three questions (and solutions) that you think will be on the final exam. Share them with your teammates.
3. Construct a monthly calendar for class work. Identify due dates for all major homework, quizzes, tests, and projects.
4. Develop a "nominal" weekly schedule that is neatly presented that includes items such as

   - Class
   - Work
   - Extracurricular activates
   - Gym
   - Study time—for each individual class
   - Other significant weekly activities Recreation time

References

1. "Studying Engineering: A Road Map to a Rewarding Career. Third Edition" Raymond B. Landis, Discovery Press, Los Angeles, CA 2007.
2. Pauk, Walter; Owens, Ross J. Q. (2010). How to Study in College (10 ed.). Boston, MA: Wadsworth. ISBN 978-1-4390-8446-5. Chapter 10: "The Cornell System: Take Effective Notes", pp. 235-277.
3. "Finding the Sum of Consecutive Numbers" Chapter 12, Lesson 5, viewed on April 1, 2019 https://study.com/academy/lesson/finding-the-sum-of-consecutive-numbers.html.
4. Stephen R. Covey, The 7 Habits of Highly Effective People, Free Press, 1989. ISBN: 0-7432-6951-9.

Chapter 4
Understanding Yourself and Others

Abstract Throughout an engineer's academic and professional career, they will have to complete both individual assignments and team projects. Therefore, it is essential for an engineer to have a good understanding of themselves, and the skills necessary to appreciate and work with people who are different from them. This chapter focuses on personal review, reflection, and growth, as well as strategies to recognize and improve inclusive and equitable practices.

By the end of this chapter, student will learn to:

- Explain Maslow's Hierarchy of Needs and how it relates to your motivation and success in college.
- Define Myers-Briggs Type Indicators and its importance in self-assessment, study habits, and teamwork.
- Discover your Myers-Briggs Type Indicator and describe how to make use of that information throughout your college career.
- Explain the strengths of diversity.
- Recognize unconscious bias and develop strategies to improve inclusive and equitable mindset and actions.
- Assess their areas of strength and weakness based on different achievement models.
- Create a personal development plan for improvement.

4.1 Understanding Yourself

Understanding yourself takes work and it is something that people work toward their entire lives. Still, having a strong understanding of oneself has many advantages including increased happiness, better decision-making, and increase tolerance and understating of others. But since human nature is complex, it takes time and work. There are many different psychological frameworks that describe human behavior. One useful framework for self-reflection is **Maslow's Hierarchy of Needs**

© Springer Nature Switzerland AG 2022
M. Blum, *An Inquiry-Based Introduction to Engineering*,
https://doi.org/10.1007/978-3-030-91471-4_4

[1]. Proposed in 1943 by American psychologist Abraham Maslow, the hierarchy shown in Fig. 4.1 correlates human's motivation with needs. According to the theory, humans are motivated to satisfy their need from the bottom of the hierarchy to the top. Once the lower-level needs are satisfied, people are more motivated to fulfill the upper levels. It is very important that lower-level needs get satisfied. This next set of questions lets us self-reflect on Maslow's Hierarchy of Needs. If you feel your basic or psychological needs are currently unsatisfied do not be afraid to reach out and seek help from university support services.

4.1.1. **Based on Fig. 4.1, explain how your physiological needs are satisfied. If this is not satisfied, explain why.**
4.1.2. **Based on Fig. 4.1, explain how your safety needs are satisfied. If this is not satisfied, explain why.**
4.1.3. **Based on Fig. 4.1, explain how your belonging needs are satisfied or not satisfied. Has coming to college made this need feel unmet?**
4.1.4. **Based on your previous answer, list some strategies to employ that will help satisfy the unmet need.**
4.1.5. **Based on Fig. 4.1, explain how your esteem needs are satisfied or not satisfied. Has coming to college made this need feel unmet?**
4.1.6. **Based on your previous answer, list some strategies to employ that will help satisfy the unmet need.**

Just as previously discussed in Chap. 2, people have different preferences. Not just for teaching and learning, but people have preferences for how they make decisions, process information, and interact with people. One of the major methods for assessing individuals is the **Myers-Briggs Type Indicator**, also known as **MBTI**. It was started in the early twentieth century by Carl Jung [2], and it was codified by Katherine Cook Briggs and her daughter Isabel Briggs Myers [3]. It is a psychometric questionnaire designed to measure psychological preferences. These types of

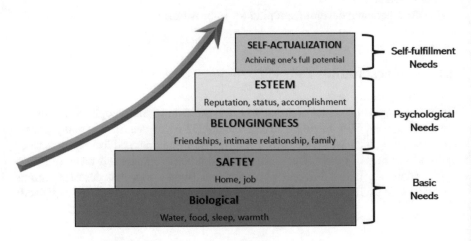

Fig. 4.1 Maslow's Hierarchy of Needs

#1: Cognitive Attitude (E vs. I)	#2: Mode of Data Gathering (S vs. N)
How a person gains energy	**How a person takes in information**
• **Extravert (E):** gains energy by activities that deal with people and things • **Introvert (I):** gains energy by activities that involve thinking about ideas and information	• **Sensing (S):** take in information with five senses. Deals with reality, with facts, tangible outcomes, and specific information • **iNtuitive (N):** take in information via impressions ands possibilities. Deals with theories or patterns

#3: Mode of Decision Making (T vs. F)	#4: Lifestyle Preferences (J vs. P)
How a person makes judgments	**How a person relates to external world**
• **Thinking (T):** uses objective principles and is concerned with logic, rules, truth and correctness • **Feeling (F):** uses subjective values and focuses more on relationships and important beliefs	• **Judging (J):** likes structures and controls in life, making decisions and sticking to them • **Perceiving (P):** likes flexibility, going with the flow, and being spontaneous

Example Tag: I S T P

Fig. 4.2 Myers-Briggs Type dichotomies of psychological preferences

questions are often used in academics to form more balanced teams and are often used by employers in hiring decisions. A Myers-Briggs profile consists of a four-letter code. Each letter is associated with a different dichotomy related to psychological preferences. Figure 4.2 explains the four dichotomies and the different preferences. With two preferences per dichotomy, there are 16 combinations that can be made. These are also categorized in terms of temperaments. Then each of the 16 personalities has a personality label and is associated with a temperament. Table 4.1 lists the 16 personalities and associated temperaments. Having a good understanding of your personality preferences can help you be cognizant of your actions and can help you improve your interactions and understanding of other people. Also, it is important to remember that these types of tests are speculative. A person's preferences can vary day-to-day, so you will be given a percentage score. Furthermore, the personality you get the first time you take the test may be different than if you take it another day, or a few years down the road, because as people grow and change, preferences change. This activity lets us explore Myers-Briggs and our personality preferences.

4.1.7. **List one way that knowing your Myers-Brigg Type can help you during a job interview.**

4.1.8. **List one way that knowing your Myers-Brigg Type can help you when you are part of a team.**

4.1.9. **Go to the following website and fill out the online questionnaire. What is your Myers-Brigg Type Profile? What is your personality label based on Table 4.1?** http://www.teamtechnology.co.uk/mmdi/questionnaire/

4.1.10. **Do you think your Myers-Brigg Type Profile is accurate? Explain Why or why not.**

Table 4.1 The 16 MBTI profiles with their aligned temperaments

Temperament	Option 1	Option 2	Option 3	Option 4
Protector (SJ constant)	ESTJ Overseer	ISTJ Examiner	ESFJ Supporter	ISFJ Defender
Creator (SP Constant)	ESTP Promoter	ISTP Crafter	ESFP Performer	ISFP Composer
Visionary (NF Constant)	ENFJ Teacher	INFJ Confidant	ENFP Advocate	INFP Dreamer
Intellectual (NT constant)	ENTJ Chief	INTJ Strategist	ENTP Inventor	INTP Engineer

4.2   Understanding Others

Personal growth will occur in a multitude of ways throughout your college career. One important area of personal growth is developing a better understanding of others, particularly people who are different from you. People are different for many reasons such as personality, gender, ethnicity, or cultural background. Today, the profession of engineering relies heavily on teamwork, and you will be successful if you can work well on a team that may differ from you.

Diversity is described as people's similarities and differences, whether they are visible or not. Research has shown that diversity is a strength in all professional contexts [4]. Diversity promotes innovation, problem-solving, and success. To increase your capacity to foster diversity and inclusion the best path to take is to educate yourself in two main areas.

The first area to educate yourself in is the area of divisive language and terms that undermine diversity and equity to mitigate those behaviors. The first term to know about prejudice is microaggressions. **Microaggressions** are actions that carry implicitly demeaning or exclusionary messages [5]. People who use microaggressions may not be aware they are using them. If a person perpetrates a microaggression without knowing and then learns about it, a poor response is a defensive response because it discourages them from taking appropriate responsibility for their actions and places the responsibility on the recipient of the aggression. The second term to know is stereotype. A **stereotype** is a widely held but fixed and simplified conception of a person or group [6].

The second area to educate yourself in is the leading way combat using stereotypes and microaggressions. **Educate yourself!** Educate yourself about people and things that are different from you. Get to know your friends, teammates, classmates, professors, and many others. Listen to them and learn about their lives and realities. Learn about new cultures, try new food, and learn about different holidays. Consciously practicing this will broaden your thinking and help you resist the urge to draw uneducated conclusions. These next activities help us begin to gain a greater understanding of ourselves and diversity, equity, and inclusion topics.

4.2.1. **Explain how diversity fosters innovation and problem-solving.**
4.2.2. **List three other benefits of having a diverse team.**

4.2.3. **For the following statements, tell whether it is a microaggression or a stereotype and identify the misconception.**

- **"There is only one race, the human race."**
- **"You are fast for a girl."**
- **"As a woman, I know what you go through as a racial minority."**
- **"I'm not racist. I have several Black friends."**
- **"I Jew-ed them down in price."**
- **"You speak very good English."**
- **"You are smart for a girl."**

4.2.4. **List elements of your identity that you think are obvious to other people.**

4.2.5. **List elements of your identity that you think are not overtly obvious to other people.**

4.2.6. **Based on the lists you made in the previous two questions, answer the following questions**

- **What elements make you feel different or unique from other people?**
- **What elements make you feel unique in a positive way?**
- **What elements make you feel different in a negative way?**

4.2.7. **Based on your answers to the previous three questions, how do you think the world sees you?**

4.3 Personal Assessment

As with many tasks in life, to accomplish them a plan must be made. The same is true for personal development and growth. To assure growth occurs, a plan must be set up. The first step is to assess yourself. Evaluating your strengths and weaknesses will allow you to develop baseline and find areas for improvement. But how do you assess yourself? What factors do you investigate? There are many different types of attributes model, but Fig. 4.3 shows three key attributes models that college students are measured against. The first model shown in Fig. 4.3a looks at the knowledge and skills a student should gain from their engineering education [7]. The second model shown in Fig. 4.3b looks at factors employers typically use to evaluate potential job candidates [8]. The last model shown in Fig. 4.3c looks at the quality of a student's education based on their level of student involvement [9]. This next activity lets us perform a personal assessment of our personal traits and college experience so far.

4.3.1. **For the engineering attributes model in Fig. 4.3a rate yourself for each attribute on a scale of 0–3, with 3 being the highest score and 0 being the lowest score.**

4.3.2. **Based on the previous analysis, what is your top area of strength?**

4.3.3. **Based on the previous analysis, what is your top area of weakness?**

Attribute	Score (0 – 3)
Able to identify, formulate, and solve complex engineering problems	
Able to produce solutions that meet specified needs with consideration of public health, safety, and welfare, as well as global, cultural, social, environmental, and economic factors	
Able to communicate effectively with a range of audiences	
Able to recognize ethical and professional responsibilities and make informed judgments, which must consider the impact of engineering solutions in global, economic, environmental, and societal contexts	
Able an ability to function effectively on a team whose members together provide leadership, create a collaborative and inclusive environment, establish goals, plan tasks, and meet objectives	
Able to develop and conduct appropriate experimentation, analyze, and interpret data, and use judgment to draw conclusions	
Able to acquire and apply new knowledge as needed, using appropriate learning strategies	

(a)

Attribute	Score (0-3)
Technical Skills - lab work, designs, software certifications, trainings	
Problem Solving and Critical Thinking – GPA, relevant coursework, academic honors	
Communication – public speaking, oral and poster presentations, written reports	
Interpersonal Skills – maturity, initiative, attitude, enthusiasm, poise, appearance, integrity, flexibility, politeness, capable to work with others and on teams	
Previous Employment and/or Leadership Experience	

(b)

Attribute	Score (0-3)
Interactions with other students	
Interactions with faculty	
Participation in student organizations	
Time spent physically on-campus	
Time spent studying for classes	

(c)

Fig. 4.3 Attribute models for assessment of self-improvement based on (**a**) engineering attributes, (**b**) job attributes, and (**c**) student involvement

4.3.4. **For the job attributes model in Fig. 4.3b rate yourself for each attribute on a scale of 0–3, with 3 being the highest score and 0 being the lowest score.**

4.3.5. **Based on the previous analysis, what is your top area of strength?**

4.3.6. **Based on the previous analysis, what is your top area of weakness?**

4.3.7. **For the student involvement in Fig. 4.3c rate yourself for each attribute on a scale of 0–3, with 3 being the highest score and 0 being the lowest score.**

4.3.8. **Based on the previous analysis, what is your top area of strength?**

4.3.9. **Based on the previous analysis, what is your top area of weakness?**

Now that you have performed a personal assessment you have a baseline for growth. For the areas that you scored high in, continue to maintain, and reinforce that strength. Then develop a plan to improve the areas that need it. This book provides many strategies and approaches for improvement and growth particularly in the previous chapters in Part I: Being an Effective Student and the subsequent Part II: The Engineering Profession. Finally, remember to reassess yourself and your skills throughout your time as an undergraduate student. This will let you track your progress and growth. The following questions help us start to build a personal improvement plan.

4.3.10. List the three areas of weakness identified in the previous activity.
4.3.11. **For each area, list two strategies you can complete this week to work to improve these skills.**
4.3.12. **For each area, list two strategies you can complete this month to work to improve these skills.**
4.3.13. **For each area, list two strategies you can complete this academic term to work to improve these skills.**

End of Chapter Questions

IBL Questions

IBL1: Answer the following questions and then discuss the answers with a partner or your team.

1. Where is your hometown?
2. Who is in your family? What are their jobs?
3. What engineering discipline(s) are you interested in studying? Why?
4. What do you consider are your strengths as a person?
5. What do you consider are your strengths as a student?
6. What do you consider are your weaknesses as a person?
7. What do you consider are your weaknesses as a student?
8. What are you most worried about coming to college?
9. What are you most excited about coming to college?

IBL2: Answer the following questions. If people feel comfortable, share with a partner or your team.

1. Make a list of the elements of your identity that you feel are visible to other people.
2. Make a list of the elements of your identity that you feel are not noticeable to other people.
3. Of the lists in the previous two questions, what elements have made you feel unique in a positive way?
4. Of the lists in questions 1 and 2, what elements have made you feel different in a negative way?
5. Of the lists in questions 1 and 2, are there any identity elements you feel you suppress or censor? Explain why.
6. Of the lists in questions 1 and 2, have you ever felt discriminated against because of part of your identity? Give an example if needed.

Practice Problems

1. Write an essay explaining your MBTI personality type. Include an explanation of the personality type, whether you agree with the results of the test, five benefits from you knowing your MBTI profile, and three ways you plan to use the information for your benefit.
2. For the following personal qualifications rate yourself on a scale of 0–3 (with 0 being the lowest score): Enthusiasm, Ambition, Initiative, Maturity, Integrity, Composure, Flexibility, Patience, Cooperation. Write an essay explaining why

you gave each rating and develop a plan to improve yourself in the three areas that were given the lowest rating.

References

1. Maslow, A.H. (1943). "A theory of human motivation". Psychological Review. 50 (4): 370–96.
2. Jung, G.G., Adler, G., and Hull, R.F.C., (1976) Psychological Types (Collected Works of C.G. Jung Volume 6), Princeton University Press, Princeton, NJ.
3. Briggs, K.C., and Myers I.B., (1977), Myer-Briggs Type Indicator, Form G, Consulting Psychologists Press, Palo Alto, CA.
4. Andrew Marder (2017) "7 Studies That Prove the Value of Diversity in the Workplace," published in Recruiting Software, https://blog.capterra.com/7-studies-that-prove-the-value-of-diversity-in-the-workplace/ viewed on May 21, 2021.
5. Sue DW (2010). Microaggressions in Everyday Life: Race, Gender, and Sexual Orientation. Wiley. pp. xvi. ISBN 978-0-470-49140-9.
6. Cardwell, Mike (1999). Dictionary of psychology. Chicago Fitzroy Dearborn. ISBN 978-1579580643.
7. Accreditation Board for Engineering and Technology (ABET), 111 Market Place, Suite 1050, Baltimore, MD 21202 (Criteria for Student Outcomes for 2021 available on webpage http://abet.org)
8. Brown, M., (2016) 5 Skills Hiring Managers Look for in Engineering Grads, viewed May 24, 2021 on https://www.engineering.com/story/5-skills-hiring-managers-look-for-in-engineering-grads.
9. Astin, Alexander W., "Involvement: The Cornerstone of Excellence," Change, July/August 1985.

Part II
The Engineering Profession

Chapter 5
The Engineering Profession

Abstract In this chapter, you will be introduced to the vast and complex world that is the engineering profession. First, we will discuss the various specializations of engineering study that a student can pursue. Followed by an investigation of the various engineering organizations that are associated with many of the different disciplines and the Grand Challenges put forth by one of the main engineering societies. Next, we investigate the different educational degrees and registrations that engineers can earn and explain the importance of accreditation in the pursuit of those degrees. Finally, we investigate the variety of employment opportunities that engineers can pursue and the daily functions of an engineering job.

By the end of this chapter, student will learn to:

- List the disciplines of engineering and explain their different areas of specialization.
- Identify your preferred engineering discipline or disciplines.
- Discover various societies associated with different engineering disciplines.
- Investigate connections and develop strategies for information gathering and engagement with relevant societies.
- Explain importance of the NAE grand challenges.
- Describe the various degree levels and career possibilities associated with an engineering education.
- Explain the ritual and significance of the Order of the Engineer.
- Describe the levels of engineering professional certifications and explain the process and importance of earning a certification.
- Understand what ABET is and its importance in engineering undergraduate programs.
- List the attributes that graduates of an ABET program must have by graduation.
- Discover various engineering job functions.
- Investigate various engineering job employment opportunities.
- Identify your current preferred job function or future career path.

© Springer Nature Switzerland AG 2022
M. Blum, *An Inquiry-Based Introduction to Engineering*,
https://doi.org/10.1007/978-3-030-91471-4_5

5.1   Engineering Disciplines

As with many job titles, there are specializations within the field. For example, within the profession of a medical doctor there are specialized areas such as pediatrics, cardiology, dermatology, and many more. So, while doctors begin their medical school journey learning similar information (anatomy, physiology, etc.), eventually they specialize in their studies and focus on topics like child health, heart health, and skin health. The same situation occurs in engineering. There are many areas of engineering specialization to choose from. Table 5.1 lists many engineering disciplines, descriptions, educational level, employment figures, and starting and mean salaries. You will find that some disciplines share common interests, require similar skills, and provide intersecting services. Some students may enter college knowing exactly the engineering discipline they want to study while others have no idea what choice they want to make. Whichever way you are it is beneficial to be knowledgeable about the variety of engineering disciplines because along the way in your academic career you may change your mind and want to switch to another major. Plus, as a professional you will have to work with engineers from different disciplines. This next activity lets us learn about different engineering disciplines and explore your current preferred discipline.

5.1.1. **Looking at Table 5.1, which engineering discipline will experience the largest job growth between 2019 and 2029?**

5.1.2. **Looking at Table 5.1, which engineering discipline will experience the largest job decrease between 2019 and 2029?**

5.1.3. **Looking at Table 5.1, which three engineering disciplines currently have the most people employed?**

5.1.4. **Looking at Table 5.1, which three engineering disciplines currently have the least people employed?**

5.1.5. **Looking at Table 5.1, which engineering discipline has the highest starting salary for current graduates?**

5.1.6. **Pick one engineering discipline you are currently interested in and write down the information for employment, job growth, starting salary, and median salary then answer the following questions regarding the engineering discipline:**

   • **Are the employment and salary numbers surprising? Explain why or why not?**
   • **Do the employment and salary numbers influence your decision to pursue this engineering discipline? Explain why or why not?**

5.1.7. **What other information are you interested in learning about the engineering discipline you choose?**

5.1.8. **Are there any other disciplines you are interested in learning more about?**

Table 5.1 Engineering Disciplines as of 2019 [1, 2]

Discipline	Description	Current employment No.	Future employment growth (2019–2029) (%)	Average starting salary	Median annual salary
Aerospace Engineering	Design primarily aircraft, spacecraft, satellites, and missiles	66,400	3	$69,041	$118,610
Agricultural Engineering	Solve problems concerning power supplies, machine efficiency, the use of structures and facilities, pollution and environmental issues, and the storage and processing of agricultural products	1700	2	$47,330	$84,410
Bioengineering and Biomedical Engineering	Combine engineering principles with sciences to design and create equipment, devices, computer systems, and software for medical applications	21,200	5	$59,982	$92,620
Chemical Engineering	Apply the principles of chemistry, biology, physics, and math to solve problems that involve the use of fuel, drugs, food, and many other products	32,600	4	$73,528	$108,540
Civil Engineering	Design, build, and supervise infrastructure projects and systems	32,900	2	$71,401	$88,570
Computer Engineering	Research, design, develop, and test computer systems and components	71,100	2	69,365	$119,560
Electrical Engineering	Design, develop, test, and supervise the manufacture of electrical equipment	328,100	3	$99,039	$103,390
Environmental Engineering	Use the principles of engineering, soil science, biology, and chemistry to develop solutions to environmental problems	55,800	3	$53,144	$92,120

(continued)

Table 5.1 (continued)

Discipline	Description	Current employment No.	Future employment growth (2019–2029) (%)	Average starting salary	Median annual salary
Health and Safety Engineering	Combine knowledge of engineering and of health and safety to develop procedures and design systems to protect people from illness and injury and property from damage	26,400	4	$72,601	$94,240
Industrial Engineering	Devise efficient systems that integrate workers, machines, materials, information, and energy to make a product or provide a service	295,800	10	$59,468	$88,950
Marine Engineering	Design, build, and maintain ships, from aircraft carriers to submarines and from sailboats to tankers	11,800	1	$65,440	$95,440
Materials Engineering	Develop, process, and test materials used to create a wide range of products	27,500	2	$61,420	$95,640
Mechanical Engineering	Design, develop, build, and test mechanical and thermal sensors and devices, including tools, engines, and machines	316,300	4	$64,501	$90,160
Mining and Geological Engineering	Design mines to remove minerals safely and efficiently for use in manufacturing and utilities	6300	4	$64,404	$93,800
Nuclear Engineering	Research and develop the processes, instruments, and systems used to derive benefits from nuclear energy and radiation	16,400	−13	$57,870	$116,140
Petroleum Engineering	Design and develop methods for extracting oil and gas from deposits below the Earth's surface	33,400	3	$77,546	$137,330

5.2   Engineering Societies

An **engineering society** is a professional organization for engineers of various disciplines [3]. The societies can be technical or civic in nature. They create a professional spirit, develop standards of practice, provided professional education, and build a sense of social responsibility among its members. Some professional societies accept many different disciplines, while others are solely discipline-specific. Many societies award honors and some award professional designations. Commonly at universities or technical colleges, there are many chapters or student-run engineering sections of the engineering societies.

Students who join these chapters of relevant societies benefit from mentoring and networking opportunities, as well as outreach and educational opportunities. A list of engineering societies and their associated websites is given in Table 5.2. Students can find more information about the various engineering disciplines from

Table 5.2   Various technical and civic engineering societies

Society name	Abbreviation	Website
Alpha Omega Epsilon	AΩE	http://www.alphaomegaepsilon.org
Alpha Pi Mu	AΠM	http://www.alphapimu.com/
American Academy of Environmental Engineers and Scientists	AAEES	http://www.aaees.org/
American Indian Science and Engineering Society	AISES	www.aises.org
American Institute of Aeronautics and Astronautics	AIAA	www.aiaa.org
American Institute of Chemical Engineers	AIChE	www.aiche.org
American Society for Engineering Education	ASEE	www.asee.org
American Society of Civil Engineers	ASCE	asce.org
American Society of Heating, Refrigerating, and Air-Conditioning Engineers	ASHRAE	www.ashrae.org
American Society of Mechanical Engineers	ASME	www.asme.org
Biomedical Engineering Society	BMES	www.bmes.org
Chi Epsilon	XE	http://www.chi-epsilon.org/
Eta Kappa Nu	HKN	www.hkn.ieee.org
Institute of Electrical and Electronics Engineers	IEEE	www.ieee.org
Institute of Industrial and Systems Engineers	IISE	http://www.iise.org/
National Academy of Engineering	NAE	nae.edu
National Society of Black Engineers	NSBE	www.nsbe.org
National Society of Professional Engineers	NSPE	nspe.org
Pi Tau Sigma	ΠTΣ	https://pitausigma.org/
Society of Automotive Engineers	SAE	www.sae.org
Society of Hispanic Professional Engineers	SHPE	https://www.shpe.org/
Society of Women Engineers	SWE	www.swe.org
Tau Beta Pi	TBΠ	http://www.tbp.org
Theta Tau	ΘT	www.thetatau.org

the society websites. These next few questions let us investigate engineering societies.

5.2.1. **List the technical society you think you would join if you were studying the following engineering disciplines**

- **Aerospace engineering**
- **Chemical engineering**
- **Civil engineering**
- **Mechanical engineering**
- **Biomedical engineering**
- **Electrical engineering**

5.2.2. **List three societies that you think would include members from multiple engineering disciplines.**

5.2.3. **List three societies that are civic in nature.**

5.2.4. **Name three societies you think you would be interested in joining and why you want to join.**

One of the societies on that list is the National Academy of Engineering (NAE). In 2008 the NAE announced its **14 Grand Challenges** for Engineering in the twenty-first century [4]. These challenges, if solved, have the potential to significantly change the way people live, work, play and improve life on the planet we live on. And each of these challenges will need the specialties of engineers to solve.

5.2.5. **Go to the website: http://www.engineeringchallenges.org/cms/challenges.aspx, list the 14 grand challenges.**

5.2.6. **Of the list created choose three challenges and explain the importance of solving them.**

5.2.7. **What grand challenge are you most interested in solving and why?**

5.2.8. **What types of engineers do you think will be needed to solve this grand challenge?**

5.3 Engineering Degrees and Registrations

As previously mentioned, the field of engineering is vast and complex. There are many opportunities for a variety of careers for someone who has an engineering degree. Also, there is a variety of collegiate degrees a person could earn in engineering. Each progressive degree indicates that you have reached a certain level of knowledge. The higher your level of degree attainment, the greater your qualifications are for certain jobs, career paths, and pay grades. Table 5.2 gives a brief description of the levels and examples of engineering professions for each degree.

It is also important to point out that earning an engineering degree means you were taught a certain way to think. You know how to solve problems. Therefore, an engineering education can be applied to almost every professional field. It is often said that *it is easier to switch to any other career from engineering than it is to switch from another career to engineering.* Students who earn a bachelorette engineering degree sometimes go on to careers in politics, policy, communications, law, medicine, finance, business, education, and many more!

5.3.1. **Based on Table 5.3, list the degrees in chronological order from first degree to final degree in engineering.**

5.3.2. **Looking at Table 5.3, explain the difference between an A.S. in Engineering and an A.S. in engineering technology.**

5.3.3. **If a student earns a B.S. in engineering and then goes straight into a Ph.D. program, nominally how long will then he/she spend in post-secondary education?**

5.3.4. **List the degree or degrees you plan to earn and explain your reasoning for wanting to reach that level of education.**

Table 5.3 Levels of degrees associated with engineering that can be earned and potential jobs that one could have with that degree

Degree	Abbreviation	Description	Years of study	Potential jobs
Associate Degree in Engineering Technology	A.S.	Prepares for a career in engineering technology. Focus on technical training. Not a licensed engineer	2	Drafting technician, manufacturing technician, maintenance technician
Associate Degree in Engineering	A.S.	Prepares for future studies and eventual earning of bachelor's degree in engineering. Focus on introductory concepts. Not a licensed engineer	2	Typically go on to earn B.S. degree
Bachelor of Science Degree	B.S.	Prepares for a career as a practicing engineer. Focus on the concept as well as practice	4	Analyst, designer, test engineer, salesperson, developer, designer, consultant
Master of Science Degree	M.S.	Research-focused = complete a thesis presenting original research in a focused engineering area Professional degrees = take classes to learn advanced techniques in more focused engineering areas	2 beyond B.S.	Manager, researcher in industry or government lab, consultant
Doctor of Philosophy	Ph.D.	The highest academic level awarded; students are required to produce original thesis research that expands the boundaries of knowledge	4–5 beyond B.S.	College professor, researcher in industry or government lab, consultant

When you earn your B.S. degree in your engineering discipline you will join the engineering profession and thus need to meet the responsibilities and characteristics of a learned professional group. Much like the medical community takes pledges based on their positions, engineering students are bound by a code of ethics (which will get discussed in detail in Chap. 8) and a code of conduct. One way to formalize this is by joining the **Order of the Engineer** [5]. Inspired by an older Canadian engineering induction ceremony [6], it is an association for graduate and professional engineers in the United States to join. It emphasizes pride and responsibility in the engineering profession. In the ceremony to join the order inductees take the Obligation of the Order and then are presented with a stainless-steel ring which is worn as a symbol of pride in the Order and obligation to the profession and the public.

Finally, as a working engineer you can earn a professional certification. The national organization that represents the state boards is the National Council of Examiners for Engineering and Surveying (NCEES) award a **Professional Engineer (PE)** license designation to those who earn it [7]. A PE license designates an engineer's proof of competency and credentials for their areas of practice. Professional Engineers have the authority to sign and seal engineering plans and offer their services to the public. To earn a PE license the following steps must be completed.

1. Earn a 4-year B.S. degree in engineering from an ABET-accredited engineering program (see next section for details about accreditation).
2. Pass the **Fundamentals of Engineering (FE) exam**—this is a 1 day 8-h multiple-choice test about the fundamentals of engineering. Students typically take the exam during their last year of undergraduate study.
3. Complete 4 years of progressive engineering experience under a licensed PE.
4. Pass the **Principles and Practice of Engineering (PE) exam**—this is another 1 day 8-h multiple-choice exam, but it is typically in the specific discipline the engineer practices in.

Finally, PE's must continually maintain and improve their skills throughout their careers to retain their licenses. In some engineering occupations, a PE license is not required. However, for engineers who work in the public sectors and may affect the health and safety of people, earning a PE license is required in the United States. These next activities help us to think about future enrollment opportunities.

5.3.5. **Go and research the Order of the Engineer. Write down the Obligation of an Engineer that is taken by each student.**

5.3.6. **Go and research the Order of the Engineer. For the Engineer's Ring that is given to each inductee, where is it worn and why is it worn there?**

5.3.7. **Which engineering disciplines do you think have the highest number of engineers who earn their PE licenses? (Think about the professionals who work in the public sector—cities, towns, governments, etc.)**

5.3.8. **Explain your current chosen career path and whether you think you will want to obtain your PE license in the future.**

5.4 ABET

When you were looking at which university or college engineering program to join you may have seen the term **ABET**. ABET use to be an acronym for Accreditation Board for Engineering and Technology. Now the acronym is the title because it has expanded beyond engineering accreditation. ABET is a nonprofit organization that provides accreditation for programs of applied science, computing, engineering, and engineering technology programs worldwide [8]. ABET consists of volunteers from academia and industry who visit campuses, review a degrees program materials, and based on criteria, decide whether a program receives ABET accreditation. Engineering programs volunteer to be evaluated for accreditation, it is not mandated. However, thousands of programs choose to be evaluated and received accreditation because it is important. When an engineering program has earned ABET accreditation it means that students who earn that degree are exposed to a certain standard level of preparation. A program which is ABET accredited demonstrates that it

- Promotes best practices in educating its students
- Directly involves college faculty and staff in self-assessment and continuous quality improvement of the program
- Educates based on "learning outcomes," rather than "teaching inputs"
- Assures that the program has received international recognition of its quality

There are many benefits for students who earn a degree from an ABET-accredited program including

- Students are assured to learn technical and professional skills that are crucial to success in the field of engineering
- Employers recognize degrees obtained from ABET programs are high quality. Many engineering positions at companies require ABET accreditation for hiring
- In order to earn a PE license, students must Earn a 4-year B.S. degree in engineering from an ABET-accredited engineering program (think back to the previous section)
- Many federal loan, grant, and scholarship programs require students to attend an ABET-accredited program
- Courses from an ABET-accredited program will transfer easier into another program

This next set of questions helps us to explore ABET.

5.4.1. **There are over 4000 programs that are accredited by ABET and they come from a range of disciplines. These disciplines fall into four groups that each have their own accreditation commission. Go to the website https://www.abet.org/ and list the four accreditation commissions.**

5.4.2. **Based on the previous question, what accreditation commission does your degree program fall under?**

5.4.3. **Is your undergraduate degree program ABET-accredited?**

5.4.4. **All ABET-accredited programs must have a set of Program Educational Objectives (PEOs). These are broad statements that describe what graduates of the program are expected to attain within a few years after graduation. If your program is accredited, list the PEOs for your program.**

5.4.5. **Each of the accreditation commissions dictates a set of Student Outcomes, also known as SOs. The SOs describe what attributes students should acquire throughout their degree program. It is the knowledge, skills, and behaviors they should gain by the time of graduation. Go to the website https://www.abet.org/ and list the Student Outcomes for your program.**

5.5   Engineering Employment Opportunities and Functions

One of the hardest perceptions for students to comprehend is what engineers do during their daily jobs. The reason it is hard to clarify is because engineering degrees lead students to so many options. Once an undergraduate engineering degree is earned students can (1) immediately seek employment as a practicing engineer or (2) they can continue with their education. And both of those choices lead to many more options. For students who seek employment the field of engineering practice provides so many opportunities. However, there are a few core job functions that engineers from any discipline can perform as a practicing engineer [9]. The core engineering job functions include

- *Design*—engineer who translates concepts and information into detailed specifications that determine the plan for development and manufacturing of something. Needs creativity and attention to detail.
- *Analysis*—engineer who mathematically models problems or product functions. Needs strong analytical ability.
- *Testing*—engineer who creates and performs tests to verify that a design or product meets specifications. Needs strong experimental skills.
- *Development*—engineer who is responsible for the development of products systems or processes. Needs technical and organizational skills to follow products from conception through design, testing, and fabrication.
- *Sales*—engineer who serves as liaison between a company and a customer. Needs strong technical and social skills.
- *Research*—engineer who uses engineering practices and scientific principles to solve applied problems that need solutions. Needs strong analytical and experimental ability typically learned through an advanced degree.
- *Management*—engineer who oversees engineering staff or technicians. Needs strong leadership skills.

- *Consulting*—Engineer who solve problems and answer questions for clients when expertise is missing. Could be self-employed or work for a consulting firm. Needs strong technical and social skills.

For a student who continues on in their education, the student either pursues graduate education (1) in a technical field or (2) nontechnical field. Technical fields include earning a master's or PhD degree in the engineering discipline they earned their undergraduate degree in or perhaps a different one. Earning a MS or PhD degree also allows engineers to go into teaching. A MS degree is typically required to teach in technical programs and community colleges and PhD degrees are required for 4-year institutions. Nontechnical fields include degrees such as medicine, law, policy, communications, business, finance, and many others. Table 5.1 displays the distribution of engineering employment for various occupational sectors [10]. This next activity lets us explore engineering job functions and employment opportunities.

5.5.1. **List your top three engineering job functions you are interested in pursuing. Explain why they appeal to you.**
5.5.2. **Based on Table 5.4, list the following occupational sectors from highest number of engineers working to lowest number of engineers working: Education, Government, Industry.**
5.5.3. **List your top occupational sector you are interested in working in. Explain why it appeals to you.**

Table 5.4   Display of sectors of employment for individuals with Engineering Degrees in 2013 [9]

Occupation	Percentage of Engineers Employed (%)
4-year institutions	5.9
2-year and precollege institutions	0.4
Federal Government	6.8
State/Local Government	6.1
Self-employed, Unincorporated business	2.6
Nonprofit Organizations	2
For-profit businesses	76.2

End of Chapter Problems

IBL Questions

IBL1:

1. Go to the website and explain what the following engineering societies are:

   - Alpha Omega Epsilon
   - Alpha Phi Mu
   - Pi Tau Sigma
   - Etta Kappa Nu
   - Chi Epsilon
   - Tau Beta Pi
   - ASTM International
   - NSBE
   - SWE
   - SHPE

2. Does your college or university have any of the societies listed above?
3. List the societies you are interested in joining.

IBL2: Answer the following questions

1. List the engineering disciplines offered by the current academic institution you attend.
2. How many students graduate annually from each discipline?
3. What is the average starting salary for a graduate from your college?
4. List three employers or job sectors that graduates of your institution work for?
5. List three graduate or professional schools that graduates of your institution attend post graduation?

Practice Problems

1. Go to the following website: https://www.youtube.com/playlist?list= PL8dPuuaLjXtO4A_tL6DLZRotxEb114cMR, choose two engineering discipline videos to watch, and write a paper explaining then comparing and contrasting the different disciplines.
2. Write a paper about one engineering society from Table 5.2. Include information such as the organization's history, mission and purpose, structure, membership eligibility and application, fellowship or scholarship opportunities, and any annual conference information.

3. Write a paper describing your top choice of engineering job function and your last choice of engineering job function. Describe and give examples of the functions, then compare and contrast the two and explain why you would like to work in one job versus the other.

References

1. Bureau of Labor and Statistics, https://www.bls.gov/ooh/architecture-and-engineering/home.htm, viewed May 26, 2021.
2. Entry Level Salaries, collegegrad.com, viewed on May 26, 2021.
3. Layton, Edwin T., Jr. The Revolt of the Engineers: Social Responsibility and the American Engineering Profession. Cleveland, Ohio: Press of Case Western Reserve University, 1971; Baltimore: Johns Hopkins University Press, 1986.
4. 14 Grand Challenges for the 21st Century, National Academy of Engineering, http://www.engineeringchallenges.org/challenges.aspx
5. Order of the Engineer, https://order-of-the-engineer.org/, viewed on May 27, 2021.
6. The Ritual of the Calling of an Engineer, https://www.ironring.ca/home-en/, viewed on May 27, 2021.
7. What is a PE? https://www.nspe.org/resources/licensure/what-pe, viewed on May 27, 2021.
8. https://www.abet.org/, viewed on June 2, 2021.
9. Landis, R.B., (2007) Studying Engineering: A Road Map to a Rewarding Career, Discovery Press, Los Angeles, California.
10. National Science Foundation, National Center for Science and Engineering Statistics, Scientists and Engineers Statistical Data System (SESTAT) (2013), http://sestat.nsf.gov.

Chapter 6
Engineering Problem-Solving

Abstract You are becoming an engineer to become a problem solver. That is why employers will hire you. Since problem-solving is an essential portion of the engineering profession, it is necessary to learn approaches that will lead to an acceptable resolution. In real-life, the problems engineers solve can vary from simple single solution problems to complex opened ended ones. Whether simple or complex, problem-solving involves knowledge, experience, and creativity. In college, you will learn prescribed processes you can follow to improve your problem-solving abilities. Also, you will be required to solve an immense amount of practice and homework problems to give you experience in problem-solving. This chapter introduces problem analysis, organization, and presentation in the context of the problems you will solve throughout your undergraduate education.By the end of this chapter, student will learn to:

- Recognize and explain the types of problems an engineer may need to solve.
- List and describe the steps in problem-solving.
- Compare problem-solving during college to problem-solving in industry.
- Explain the importance of properly presenting the solution to a problem.
- List the components of a properly presented solution.
- Apply the problem-solving technique and the guidelines for proper presentation to solve and present problems in a systematic and logical manner.

6.1 Types of Problems

A **problem** is defined as a situation, faced by an individual or a group of individuals, for which there is no obvious solution [1]. An engineer's education is focused on improving one's ability to think of solutions logically and creatively to problems. There are many **types of problems** an engineer may encounter such as [2, 3]

- **Research**—Problems that require a hypothesis to be proved or disproved.
- **Knowledge**—Problems when someone encounters a situation they do not understand and further investigation is needed.

© Springer Nature Switzerland AG 2022
M. Blum, *An Inquiry-Based Introduction to Engineering*,
https://doi.org/10.1007/978-3-030-91471-4_6

- **Troubleshooting**—Problems when equipment behaves in unexpected or improper ways.
- **Mathematics**—Problems that need to describe physical phenomena with mathematical models.
- **Resource**—Problems are encountered due to the limitation of supply of certain things.
- **Societal**—Problems that affect or influence many citizens within a society.
- **Personal**—Problems that are privately experienced within the character of an individual and within the range of their immediate relation to others.
- **Design problems** are the heart of engineering. Solving them requires creativity, teamwork, and broad knowledge.

This next activity looks at the types of problems engineers will encounter.

6.1.1. **List three types of problems that are the most frequently encountered by engineering students. Explain your reasons for your list.**

6.1.2. **List three types of problems that are the most frequently encountered by engineering professionals. Explain your reasons for your list.**

6.1.3. **Give an example of a societal problem an engineer might encounter.**

6.1.4. **Give an example of a resource problem an engineer might encounter.**

6.1.5. **Give an example of a personal problem an engineer might encounter.**

6.1.6. **Do engineers need to be concerned with societal problems; Yes or No? Explain your answer and use examples of societal problems.**

6.2   Problem-Solving Technique

Being a good problem solver is a defining characteristic of an engineer [2, 3]. Problem-solving involves a combination of *knowledge* and *skill*. The *knowledge* needed includes understanding principles of physics, chemistry, mathematics, and other subjects like mechanics, thermodynamics, and fluids. The *skill* involved includes using proper judgment, logic, experience, and common sense to appropriately apply the knowledge. While problem-solving in college may vary slightly from problem-solving in a professional capacity, the steps involved are the same, namely

1. Identify the problem
2. Gather Data/Information
3. Analyze using applicable principles
4. Find Solution
5. Verify Solution

Figure 6.1 details guidelines for problem-solving and divides details of each process based on how it is typically applied in college versus professional real-life situations.

Fig. 6.1 Problem-Solving Technique used in both academia and industry

IDENTIFY THE PROBLEM

School	Industry
• Professor gives you problem statement	• Identified by manager or client

GATHER INFORMATION/DATA

School	Industry
• Read problem statement and identify pertinent information	• Identified and gathered by engineer or project team

ANALYZE USING APPLICABLE PRINCIPLES

School	Industry
• Use theories, constraints, assumptions from class/lecture	• Use scientific principles, assumptions from learned knowledge and experience

SOLVE PROBLEM

School	Industry
• Solve mathematical models using freehand, calculator or computer program	• Solve mathematical models using freehand, calculator or computer program

VERIFY SOLUTION

School	Industry
• Check professors or textbook results to see if the mathematical answer makes physical sense and is reasonable	• Check results are within magnitude bounds and problem statement specifications

6.2.1. **For the following problems identified, label whether they most likely came from school or industry.**

- **A 5500-lb truck is lifting a 2000-lb boulder that is on a 200-lb pallet. Knowing that the acceleration of the truck is 2 ft/s² determine the force between the truck tires and the ground.**
- **Design a car that goes from 0 to 70 miles per hour in 5.0 s, gets 30 miles-per-gallon fuel economy, costs less than $50,000, meets government safety and pollution standards, and appeals to the aesthetic tastes of people 18–25 years old.**

6.2.2. **Looking at Fig. 6.1, which step in the problem-solving process is the same for both school and industry?**

6.2.3. **Looking at Fig. 6.1, explain the difference between Step 3—Analyze using applicable principles—in industry versus school.**

6.2.4. **If you compute the monthly payment for a school loan of $7,000 dollars over a 5-year period at an annual interest rate of 8%, and you arrive at an answer that you need to pay $10,000 dollars per month, is this a reasonable solution?**

6.2.5. **Which step in the problem-solving technique would the previous question fall under?**

6.3   Problem-Solving Presentation

Once an engineer has solved a problem, they must communicate the solution to others. It is important that they communicate in a precise, logical, and clear manner so that others can also understand the solution. Engineering communication for written reports, proposals, posters, and oral presentations will be discussed in detail in Chap. 10 in engineering communication. This section will focus on how to present "closed-ended solutions," because engineering students typically see many of these types of problems throughout their engineering education [2, 3]. Following these general guidelines allows the student to effectively document and communicate the method of solution. This allows for easy double-checking of work for correctness and allows for faculty and graders to be effective in locating mistakes and providing useful feedback to students. Some engineering schools may mandate a homework format, while others may not. Below are general steps for the appearance and documentation of an engineering solution.

Step 1: Create Problem Statement

- State the question to be solved in a concise manner.
- Identify all known and unknown information using symbolic notation. It is good practice to mark unknown with question marks.
- State any assumptions that are being made if they are needed. Sometimes assumptions will be stated in the problem statement, sometimes they will not.

- Draw a diagram of the situation if appropriate. Make sure to label any relevant data on the diagram.

Step 2: Identify the theory, principles, or relationships being used

- Write out the main equations that contain desired quantity. Sometimes the equations will need to be derived.
- Write or derive any complimentary equations or relationships that may need to be used along with the main equation.

Step 3: Solve the problem

- Isolate the desired quantity. This typically requires algebraic manipulation.
- Write out a description of each step you take. Make sure to not just write the equations.
- Sometimes it is easier to insert numerical values after all algebraic manipulation and substitutions are made. Sometimes calculations will have to occur throughout the solution.
- Ensure that units cancel appropriately and check for sign errors.
- Compute the answer.

Step 4: Identify results and verify the answer

- Clearly identify the answer to the question. Sometimes an underline or a box around the answer is required. Make sure to include the units!
- Check that the final answer makes physical sense, is realistic and accurate.
- Confirm that all questions have been answered.

Other important elements to include in your paper

1. Your Name
2. Date the work was completed
3. Course and section number
4. Homework Assignment number
5. Page number

Finally, some other insights include

- Remember to check your final units for sense and scale!!
- It is recommended to perform work on engineering paper. It is writing paper that is printed with fine lines making up a regular grid. The grid is visible on the front and faint on the back so that people only write on the front of the page. It typically comes in pads of white or green.
- It is recommended that calculations be performed in pencil because errors will be made, and work will have to be redone.
- Solution should be as neat as possible so it can be easily checked by another person.

When students execute these steps properly it leads to a better understanding of problems. Also, in the future, it will allow developing engineers to solve and document a wider range of more complex problems. This next activity lets us begin to develop our problem-solving and presentation abilities.

(a)   (b)

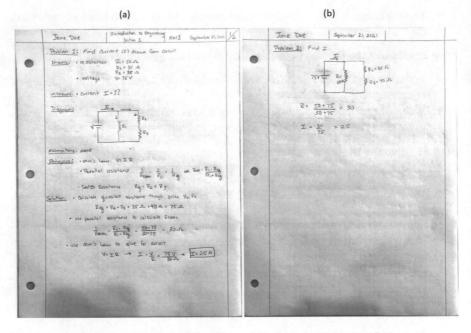

Fig. 6.2 Sample free-hand homework problems. (**a**) Example 1 and (**b**) example 2

6.3.1. **Why do you think some schools mandate that all homework is completed on engineering graph paper?**

6.3.2. **Based on Fig. 6.2, which homework assignment is properly formatted?**

6.3.3. **Based on Fig. 6.2, how many pages is the homework shown in Fig. 6.2a?**

6.3.4. **Based on Fig. 6.2, explain all the incorrect items about the improperly formatted homework.**

6.3.5. **Answer the following question using the steps for appearance and documentation of an engineering solution: An object is in static equilibrium when all the moments balance. A moment is a force exerted at a distance from a fulcrum point. Also recall a person's weight is their mass times the acceleration due to gravity. A 35 kg child and another child sit on a 6-m long seesaw. Neglecting the weight of the beam, what does the mass of the second child have to be for the beam to be balanced if the fulcrum is placed 2 m from the 35 kg child?**

End of Chapter Problems

IBL Questions

IBL1: Using standard problem-solving technique, answer the following questions

1. If you run in a straight line at a velocity of 10 mph in a direction of 35 degree North of East, draw the vector representation of your path (hint: use a compass legend to help create your coordinate system)
2. If you run in a straight line at a velocity of 10 mph in a direction of 35 degree North of East, explain how to calculate the velocity you ran in the north direction.
3. If you run in a straight line at a velocity of 10 mph in a direction of 35 degree North of East, explain how to calculate the velocity you ran in the east direction.
4. If you run in a straight line at a velocity of 10 mph in a direction of 35 degree North of East, explain how to calculate how far you ran in the north direction.
5. If you run in a straight line at a velocity of 10 mph in a direction of 35 degree North of East, explain how to calculate how far you ran in the east direction.
6. If you run in a straight line at a velocity of 10 mph in a direction of 35 degree North of East, how far north have you traveled in 5 min?
7. If you run in a straight line at a velocity of 10 mph in a direction of 35 degree North of East, how far east have you traveled in 5 min?
8. What type of problem did you solve?

IBL2: For the following scenarios, explain what type of problem it is that needs to be solved.

- Scientists hypothesize that PFAS chemicals in lawn care products are leading to an increase in toxic algae blooms in lakes during summer weather.
- An engineer notices that a manufacturing machine motor hums every time the fluorescent floor lights are turned on.
- The U.N. warns that food production must be increased by 60% by 2050 to keep up with population growth demand.
- Engineers are working to identify and create viable alternative energy sources to combat climate change.

Practice Problems

Make sure all problems are written up using appropriate problem-solving technique and presentation.

1. The principle of conservation of energy states that the sum of the kinetic energy and potential energy of the initial and final states of an object is the same. If an engineering student was riding in a 200 kg roller coaster car that started from rest

at 10 m above the ground, what is the velocity of the car when it drops to 2.5 m above the ground?

2. Archimedes' principle states that the total mass of a floating object equals the mass of the fluid displaced by the object. A 45 cm cylindrical buoy is floating vertically in the water. If the water density is 1.00 g/cm3 and the buoy plastic has a density of 0.92 g/cm3 determine the length of the buoy that is not submerged underwater.

3. A student throws their textbook off a bridge that is 30 ft high. How long would it take before the book hits the ground?

References

1. https://www.merriam-webster.com/dictionary, viewed June 3, 2021.
2. Mark Thomas Holtzapple, W. Dan Reece (2000), Foundations of Engineering, McGraw-Hill, New York, New York, ISBN:978-0-07-029706-7.
3. Aide, A.R., Jenison R.D., Mickelson, S.K., Northup, L.L., Engineering Fundamentals and Problem Solving, McGraw-Hill, New York, NY, ISBN: 978-0-07-338591-4.

Chapter 7
Engineering Design and Experiments

Abstract The word design has many meanings and encompasses a wide range of professions. Professions such as fashion designer, graphic designer, interior designer, and advertising designer all have the word design in it. However, the previously mentioned professions are focused on appearance and esthetics. Engineering design refers to the processes, techniques, and scientific principles that are implemented to bring about the realization of a product, device, system, process, or experiment. This chapter will explore the fundamental steps that engineers follow in designing products or when designing an experimental test to be performed.
By the end of this chapter, students will learn to:

- Recognize and explain the key steps in the design process.
- Explain the importance of the engineering design process in the ability to produce solutions that meet society's needs.
- Recognize and explain the key steps in designing an experiment.
- Translate a research question into a designed experiment.

7.1   The Engineering Design Process

The engineering design process is a method for finding solutions to the problems faced by humanity. The systematic design process allows engineers to design new products, systems, and processes or to improve on old ones. In today's technologically complex society, many problems are now solved by teams of engineers from different disciplines. While there are many variations among the details [1–4], there are fundamental steps that every engineering discipline follows in the design process.

Step 1: Identify and define the problem. Problems can be identified and defined based on a new idea, consumer needs, societal issues, or defined by a company.

Step 2: Gather information. This step requires the engineer to perform investigation on the background of the problem, so they fully understand it. This typically

M. Blum, *An Inquiry-Based Introduction to Engineering*,
https://doi.org/10.1007/978-3-030-91471-4_7

involves patent searches, acquiring preliminary data through experiments or reading test result, internet searches, journal articles, company documents, etc.

Step 3: Identify constraints and impacts. Once the scope of the problem is understood, the limitations of the solution can be found. Factors such as material, cost, time, amount, accessibility, environmental, cultural, and social may result in design restrictions which need to be accounted for.

Step 4: Devise concepts and ideas for solution. Many people consider this the brainstorming step. Engineers want to generate many alternative solutions. The more creative and the more variety of solutions, typically the more successful the design will be. This step becomes easier with experience and knowledge.

Step 5: Evaluate the concepts to narrow down solutions. Each of the alternative solutions is evaluated. Depending on the situation, evaluation may include technical and computer analysis or prototyping and testing to reduce the choices to a final workable solution. This step becomes easier with experience and knowledge.

Step 6: Create, analyze, and test the final solution. In this step, the final solution is fleshed out in detail. Substance is given to the design by creating the final architecture and components and material choices. Final theoretical or experimental analysis is performed to prove that the design fulfills the problem statement.

Step 7: Communication and presentation of results. This requires the dissemination of the information to parties outside of the design team. Typically, it involves final reports, posters, or presentations to management and customers.

Throughout all the previous steps described, engineers should be documenting and keeping organized records of their work. This includes team meeting agendas and minutes, design notebooks, and tracking their time allocated to the project. Be aware that although engineering design is an ordered approach with specific steps, it is not a linear process. Proceeding straight from one step to the next will not necessarily reach the best solution. It is an iterative process. New knowledge and additional data may require the engineer to rework previous steps. Throughout a student's engineering education, they will study and focus on pieces of the overall design process. For example, much of the technical knowledge you acquire throughout your classes will be applied in steps 5 and 6. However, most senior engineering students will participate in a culminating capstone design experience that encompasses the whole engineering design process and will test their accumulated engineering knowledge from their degree classes.

7.1.1. **Give an example of a place in the design steps where an engineer might have to reiterate and rework previous steps.**

7.1.2. **Look up the following two product design initiatives: Product Life Cycle Management, Design for Environment. (1) explain what is meant by the terms and (2) give an example of how they are currently being applied during the design process.**

7.1.3. **Explain the importance of the engineering design process in the ability to produce solutions that meet society's needs with consideration of public health, safety, welfare, as well as global, cultural, social, environmental, and economic factors.**

7.1.4. **Define a problem to be solved, or a product, device, or services that need to be created.**

7.1.5. **From the problem you have defined, list all the pertinent information you will need to find to understand the problem.**

7.1.6. **For the problem you have defined, brainstorm as many ideas for solutions as you can.**

7.2 Experimental Design Process

While in college you will learn how to solve many problems using useful theories and equations. However, these theories typically model simplified versions of physical phenomenon. The fundamental equations tend to neglect numerous factors that affect the precision of the answer. Since it is difficult to accurately predict behavior analytically, engineers also must become skilled at conducting experiments.

An experiment is a procedure carried out to test a proposed explanation of a problem. It is a major component of the scientific method [4, 5]. There are a variety of ways the scientific method is presented, but the essential steps include

- Make an observation and ask a question.
- Form a hypothesis or testable explanation or solution for the question.
- Create a prediction of behavior based on the hypothesis.
- Test the prediction through well-designed repeatable experiments.
- Communicate results and use the results to draw conclusions and make new hypotheses or predictions.

To run a useful and productive experiment, an organized and coherent plan must be created. Again, there are various approaches to creating a good experiment [5, 6], but listed below are common standard steps and guidelines.

Step 1: Define your research question and hypothesis. First adequate background study needs to be completed to become well versed in the field of study and the problem. Based on the background research, develop a specific research question and hypothesis for solution to the question.

Step 2: Define your experimental variables. To translate a research question and hypothesis into testable form, the main variables need to be defined and relationships between variables need to be established. In an experiment, you manipulate one or more *independent variables* and measure their effect on one or more *dependent variables* (for more details on variables see Chap. 18 on Mathematics). A good practice is making a table of the research question, independent and dependent variables. The extraneous variables also need to be identified. These are the variables that are not being investigated but can potentially affect the outcomes of the research study. It is good practice to make a table of the research question, the possible extraneous variables, and a way to control or account for them.

Step 3: Establish experimental protocols. Once the variables are all identified the procedures can be developed. This will vary based on the types of experiments being conducted but important items to address are

- Establish the range and increments the independent variables will be tested over.
- Determine the number of times the experiment will be repeated at each changing variable increment. A good rule of thumb is to conduct the experiment at least three times at each increment. This will allow for a minimal statistical analysis to be run. Also, establish an order for varying the parameters.
- Establish the types of equipment needed to control the experiment. Also, identify the method for collecting and analyzing data.

Once the experimental protocol has been established and well documented, the experiments can be run, the data can be collected, analyzed, and conclusions can be drawn and potentially used to generate new hypotheses or predictions. This next activity introduces us to experimental design.

Scenario: You are interested in how the type of ball affects speed when rolled on a significant slope. Specifically, you ask how fast a tennis ball, lacrosse ball, and a ping-pong ball can roll down a ramp.

7.2.1. **For the given scenario, is there a way you can theoretically model this question?**
7.2.2. **Based on the previous question, list three limitations of the theoretical model?**
7.2.3. **For the given scenario, fill in Table 7.1.**
7.2.4. **For the given scenario, fill in Table 7.2.**
7.2.5. **For the given scenario, develop an experimental protocol to test your question, specifically establish:**

- **The range the independent variables will be tested over**
- **The increments the independent variables will be tested over**
- **The number of times the experiment will be repeated at each increment**
- **The order for testing the parameters**
- **The equipment needed for the experiment**
- **The method for collecting data**
- **The method for analyzing data**

Table 7.1 Independent and dependent variables for research question

Research question	Independent variables	Dependent variable
Type of Ball versus Speed		

Table 7.2 Extraneous variables and controls for research question

Research question	Extraneous variables	How to control
Type of Ball versus Speed		

End of Chapter Problems

IBL Questions

IBL1: For the scenario, answer the following questions. Scenario: You are interested in determining the coefficient of friction for four different materials; rubber, silk, metal, and plastic.

1. What is your research question and hypothesis?
2. What are the independent, dependent, and extraneous variables?
3. How do you plan to control the extraneous variables?
4. What is your experimental protocol to test your question, specifically establish:

   - the range the independent variables will be tested over
   - the increments the independent variables will be tested over
   - the number of times the experiment will be repeated at each increment
   - the order for testing the parameters
   - the equipment needed for the experiment.
   - the method for collecting data
   - the method for analyzing data

Practice Problems

1. Use the engineering design process to design one of the following items:

   - Prosthetic limb—leg for a dog
   - Human-powered water purification system
   - Sorting Machine that can be used to separate and package nuts, bolts, and washers

2. Use the experimental design process to design an experiment for the following:

   - Determine the density of bone, glass, and concrete and compare with published material property results.
   - Determine the heat capacity for bone, glass, and concrete and compare with published material property results.

References

1. Cross, Nigel (2000), Engineering Design Methods: Strategies for Product Design (3rd Edition), John Wiley & Sons, West Sussex, England.
2. Aide, A.R., Jenison R.D., Mickelson, S.K., Northup, L.L., Engineering Fundamentals and Problem Solving, McGraw-Hill, New York, NY, ISBN: 978-0-07-338591-4.
3. Saeed Moaveni (2005). Engineering Fundamentals: An Introduction to Engineering (2nd ed.). Thompson. ISBN: 0-534-42459-7.

4. Stephan, E.A., Bowman, D.R., Park W.J., Sill B.L., Ohland, M.W. (2018), Thinking Like an Engineer: An Active Learning Approach, 4th Edition, Pearson, Upper Saddle River, New Jersey.
5. scientific method, Oxford Dictionaries: British and World English, 2016, viewed on June 9, 2021.
6. Rebecca Bevans (2021), A guide to experimental design, https://www.scribbr.com/methodology/experimental-design/, viewed on June 9, 2021.

Chapter 8
Engineering Ethics

Abstract As a member of humanity, you make decisions every day. The choices you make shape your life's journey and your reputation. As an aspiring engineer, the decisions you make effect your career and potentially the health, safety, and lives of other people because engineers build and design many products and services that affect others. Therefore, engineers need to have moral decision-making capability and adhere to high ethical standards. Ethics provides a moral philosophy and a set of conduct standards that helps people decide how to act in situations. This chapter discusses the importance of engineering ethics and teaches how to practice moral decision-making.

By the end of this Chapter, students will learn to:

- Explain the steps in the ethical decision-making process.
- Discuss and practice the ethical decision-making process.
- Recognize violations of ethics and academic integrity that can occur as a student.
- Identify and describe the code of conduct and academic integrity policy of their college institution.
- Recognize violations of ethics and academic integrity that can occur as a practicing engineer.
- Identify and describe the NSPE code of Ethics.

8.1 Making Ethical Decisions

Some ethical choices are straightforward and universally understood. However, many decisions people must make are not easy or obvious. There are many considerations, views, and opinions when making decisions so there is no set formula to guarantee that the most ethical decision gets made. However, there are processes and steps that can be followed when judging problems with ethical consequences. Several ethical decision-making models are presented in the literature [1, 2], and while they vary in number of steps, they are similar in design and content. Several standard steps are provided in the models for making choices [2].

© Springer Nature Switzerland AG 2022
M. Blum, *An Inquiry-Based Introduction to Engineering*,
https://doi.org/10.1007/978-3-030-91471-4_8

1. **Identify the Ethical Dilemma**. Clearly and succinctly state the ethical problem and consider all pertinent aspects.
2. **Information Gather**. Collect information, data, and facts about the situation. Identify the stakeholders in the decision and who would be effected by the decision.
3. **Identify the Options**. Brainstorm and list all the outcomes that are possible based on the choice that is made.
4. **Apply Ethical Principles, Codes, and Beliefs**. Identify relevant codes of ethics and guidelines, and reflect on personal principles and values. Apply each of these filters to the problem and see how the decision is affected.
5. **Make and Implement a Decision**. Act on the decision that has been made.

This process is dynamic and Steps 3 and 4 will vary from person to person and take the most work. However, following a guided process will allow for more educated, deliberate principled decision-making. These next questions practice the process of ethical decision-making

8.1.1. **Give an example of a universally understood ethical decision.**
8.1.2. **For the following questions answer yes or no.**

  - **Do you knowingly violate the speed limit when you are driving in a car?**
  - **If a student has cheated on an assignment and you know about it, do you turn the student in?**
  - **If you see a homeless person begging for handouts, do you give them something?**
  - **A friend you are driving with runs a red light, do you say something to them?**
  - **You go to a party and your underage friend is drinking, do you say something to them or ask them to stop?**
  - **Would you download music illegally off the internet?**

8.1.3. **For the questions answered above, go through the ethical decision-making process and discuss why you answered that way.**

8.2   Ethics as a Student

When you choose your college, you became a member of that community. And with that privilege comes responsibilities. As a student there are ethical expectations for their behavior both in the classroom and out in the campus community. Universities will have a **code of conduct** which all students are expected to abide by. This explains actions that will result in disciplinary measures. The next activity explores violations of ethical student behavior.

8.2.1. **Where can you find your institution's code of conduct? (e.g., website, receive a book at orientation, etc.)**

8.2.2. **Based on the institution's code of conduct, what is the disciplinary measure for the following actions [3]:**

- **Theft of or damage to the university, personal, public, or private property/services or illegal possession or use of the same.**
- **Forgery, alteration, or fabrication of identification cards, records, reports, grades, diplomas, university documents, possession or purchase of falsified identification cards, or misrepresentation of any kind to a university office, university official, or law enforcement.**
- **Unauthorized entry or use of university facilities that are locked, closed, or otherwise restricted as to use.**
- **Disorderly conduct including, but not limited to, public intoxication, lewd, indecent, or obscene behavior.**
- **Illegal use, possession, purchase, distribution, manufacture, or sale of alcohol, drugs, or drug paraphernalia.**
- **Unauthorized possession or use of any weapon.**
- **Physical harm or threat of physical harm to any person or persons, including, but not limited to: assault, sexual abuse, or other forms of physical abuse.**

While the student code of conduct addresses many ethical behaviors, one that students violate most often is academic integrity. Therefore, there is typically a separate policy that addresses academic integrity. **Academic integrity** refers to college institution's expectations that students demonstrate honest and moral behavior in an academic setting. This includes areas of teaching, learning, research. While there are many specific forms academic dishonesty can take, there are some broad classifications to be aware of [4, 5]:

- **Plagiarism**—refers to when someone intentionally and knowingly represents ideas, words, and work as their own without acknowledging the source.
- **Cheating and Contract Cheating**—refers to using or attempting to use unauthorized information, materials, devices, sources, or practices in completing academic activities.
- **Fabrication and Falsification**—refers to the unapproved creation or alteration of information in an academic document or activity.
- **Sabotage**—refers to disrupting or destroying another person's work so that they cannot complete an academic activity successfully.

Whether intentional or unintentional, there are consequences for violating an institution's academic integrity policy. The next activity explores violations of academic integrity.

8.2.3. **Where can you find your institution's academic integrity policy? (e.g., website, receive a book at orientation, etc.)**

8.2.4. **For each of the following listed, identify which classification of academic integrity it falls under:**

- **Copying and pasting material from a website into your own document without proper citation.**

- Failure to contribute as required to a team project.
- Artificially creating data when it should be collected from an actual experiment.
- Faking illness or telling lies to avoid taking an exam at a scheduled time.
- Working from the textbook or note sheet during a closed book exam.
- A student who allows another student to copy from his or her work.
- Making up a source of information that does not exist.

8.2.5. Based on the institution's academic integrity policy, what is the disciplinary measure for the previous actions listed above?

8.2.6. Give another example of plagiarism violations not listed above.

8.2.7. Give another example of cheating violations not listed above.

8.3   Ethics as an Engineer

Ethical decision-making does not stop when college is over. As a practicing engineer, there will be choices made that effect the safety and well-being of many different people. Humanity trusts engineers to have specialized knowledge and skills that are being used for honest, helpful, and good purposes. If those knowledge and skills are used corruptly, catastrophes and tragedies can occur. Therefore, many engineering societies have developed their own codes of ethics for conduct in their specific professions. Generally though, they are based on the National Society of Professional Engineers (NSPE) Code of Ethics [6]. The **NSPE Code of Ethics** is a guide for ethical dilemmas for practicing and professional engineers. It consists of the preamble, the fundamental cannons, rules of practice, professional obligations, and the engineer's creed. The next activity introduces ethics in engineering practice.

8.3.1. Go to the website: https://www.nspe.org/resources/ethics/code-ethics. Write down the fundamental canons for the NSPE code of ethics.

8.3.2. Go to the website: https://www.nspe.org/resources/ethics/code-ethics. How many rules of practice are there?

8.3.3. Go to the website: https://www.nspe.org/resources/ethics/code-ethics. How many rules of professional obligations are there?

8.3.4. For each of the following scenarios, state whether you think it is ethical or unethical. If you are unsure, you can state that also.

- Using a fake ID to purchase alcohol.
- Taking time at a summer job to study for summer class.
- Using the company copy machine to print up a package return label.
- Accepting an invitation to lunch from a sales representative trying to sell a new machine to your company.
- Taking home a company computer monitor.

8.3.5. Explain your answer for each scenario in the previous question.

End of Chapter Problems

IBL Questions

IBL1: Use the five-step process discussed in Sect. 8.1 and write up how you would handle the following social situations.

- Your roommate has a fake ID to purchase alcohol and has asked you to contribute money to buy a keg for a party being thrown at your residence.
- You saw your best friend's boyfriend having lunch with his ex-girlfriend. The next day your best friend says her boyfriend said he went out to lunch with his teammates.

IBL2: Use the five-step process discussed in Sect. 8.1 and write up how you would handle the following academic situations.

- A few friends from your math class have purchased a subscription to Chegg.com and have been using it to find answers to the homework problems, and then submitting them as their own work.
- You see that your professor has recently mis-scored your test. They gave you a score of 76 when it should really be a 66.
- Your friend has been sick, and they ask to copy your homework.

IBL3: Use the five-step process discussed in Sect. 8.1 and write up how you would handle the following professional situations.

- A material supplier is trying to sell raw plastic to your manufacturing firm and they offer to buy you lunch.
- A material supplier is trying to sell raw plastic to your manufacturing firm and they offer to buy you a new car.
- At work, you see a coworker using the company copy machine to make copies of a bake sale flier for their children's school.

Practice Problems

1. Write a paper reflection on the NSPE Code of Ethics preamble and the Engineer's creed. What do the code and creed mean to you and how will you employ it throughout your career?
2. Write a paper about the engineering ethics of a chosen technology that is important or interesting to you. Make sure to include the following components:
   - Describe the technology and list five intended and five unintended effects of this technology.
   - Explain how the unintended effects affect users or nonusers.
   - Explain how the unintended effects affect users or nonusers.

- Explain if the benefits of the intended effects outweigh the unintended effects.
- Discuss the role that engineers played in the development of this technology. Make sure to address specific engineering societies and their Codes of Ethics.
- Discuss the role that engineers can play to eliminate (or decrease) the unintended effects.

References

1. Deborah G. Johnson (2020), Engineering Ethics, Yale University Press, ISBN: 978-0-300-20924-2.
2. Beauchamp TL, Childress J. Principles of Biomedical Ethics 7th edition. New York. Oxford University Press. 2012.
3. Syracuse University Student Code of Conduct, https://policies.syr.edu/policies/academic-rules-student-responsibilities-and-services/code-of-student-conduct/, viewed June 11, 2021.
4. Bernard E. Whitley Jr. & Patricia Keith-Spiegel (2001) Academic Integrity as an Institutional Issue, Ethics & Behavior, 11:3, 325-342, DOI: https://doi.org/10.1207/S15327019EB1103_9
5. Gregory J Cizek (2003), Detecting and Preventing Classroom Cheating: Promoting Integrity in Assessment, Corwin Press Inc., Thousand Oaks, California. ISBN: 0-7619-4655-1
6. NSPE Code of Ethics, **https://www.nspe.org/resources/ethics/code-ethics,** viewed on June 11, 2021.

Chapter 9
Engineering Communications

Abstract In academia, engineers spend much of their time studying and work on mathematics, science, and problem-solving. However, communication is an integral part of engineering work. Successful engineers have both technical and communication skills. As an engineer, it is very important to have good communication skills because complex ideas can be effectively relayed to coworkers or superiors which will lead to success at one's jobs. Effective communicators are also often seen as strong leaders. This chapter will discuss guidance and practices for developing effective oral, written, and team communication skills.

By the end of this Chapter, students will learn to:

- Explain the general guidelines and format for writing technical documents.
- Compare and contrast the style and formats for technical design reports and laboratory reports.
- Identify common mistakes in technical writing.
- Identify and compare the variety of types of engineering documents.
- Apply best practices for graphical presentation of data in the forms of tables, figures, and graphs.
- Examine effective techniques for giving a slide show and poster presentation.
- Explore best practices for giving oral presentations.
- Explore strategies and practices for communicating and working effectively in a team.

9.1 Written Communication 1: Technical Writing

The main staple of engineering practice is the pursuit of value. Engineers work to solve problems and develop or improve products, devices, systems, and methods that others will benefit from or find value in. Finding value in the work is what breeds passion for working. Though, most times engineers struggle with written communication. However, written communication is incredibly valuable to the

© Springer Nature Switzerland AG 2022
M. Blum, *An Inquiry-Based Introduction to Engineering*,
https://doi.org/10.1007/978-3-030-91471-4_9

success of engineering pursuits particularly because it is one of the main forms for articulating the value of the work. Every piece of writing an engineer creates has a purpose. And the objective often dictates the structure that the written communication takes. In many instances, it is necessary for an engineer to produce technical written communication. **Technical communication** is defined as exhibiting characteristics such as communicating about technical or specialized topics, applications, procedures, or regulations or providing instructions about how to do something [1]. Basic forms of technical writing include user manuals, installation guides, and operating procedures. Two forms of technical writing that engineers frequently create are design and laboratory reports.

A **laboratory (lab) report** is a written document that describes what an engineer *did* and *learned* while performing a hands-on experiment. While in school, a lab report is based on assignments performed during class. As practicing engineer, experiments may be performed as required by the job position. Regardless of whether written for school or work, lab reports provide written evidence of insight into what happened, why it happened, and what it means in relation to the experimental aims. While lab reports can vary in length and format, most are structured logically in accordance with similar conventions and basic structures [2–4]. The standard parts of a lab report are described below, a long with some best practices for each section.

Title—this is the name of the report. It should clearly describe the work presented and it should draw the attention of the reader.

Tips: Avoid using determiners such as "the," "a," or "an" as the first word in a title because it leads to vague searches if someone tries to search for the work, such as in a database.

Abstract—Sometimes called the summary, this section gives an overview of the entire report. It should tell the reader (1) the aim of the experiment, (2) a brief description of the method(s) used in the experiment, (3) the results found and the interpretation of the results and (4) what the results mean in relation to the aim of the experiment.

Tips: Do not include background information such as history or theory in the abstract, because it crowds the abstract and it will be covered in the introduction. When writing an abstract, make sure the aim or aims are clearly written (the first aim is…). Also, it is recommended sometimes to write the abstract last because then pieces can be pulled from the completed lab text.

Introduction—Sometimes called the background, this introduces the topic and reason for performing the experiment. Background information can include relevant laws, theories, equations, assumptions or figures necessary, and any previous analysis or experiments pertaining to the question being answered by the experiment.

Tips: Always cite sources for the introduction including lecture notes, discussions with professor, or lab handouts.

Method—Also called the procedure, this section describes step-by-step how the experiment was performed. Information needed to be included is (1) description of materials and apparatus used (2) data collection method (3) result calculation method (4) result analysis method, and (5) any difficulties that were encountered and how they were overcome or impacted the study.

Tips: Provide an image of the experimental setup with clearly labeled components or a flowchart of the testing process with relevant variables and times clearly labeled. Also, include enough detail for another person to replicate the experiment and get a similar outcome.

Results and Analysis—This section shows the final computations and data that was collected and analyzed from the experiment. The results are typically displayed in graphs, figures, or tables (see Sect. 9.3 for details about suitable engineering graphics) and their significance to the experimental aim is explained.

Tips: Include an explaination of how the raw data was processed to obtain the final results and include an error analysis where applicable. Also, present your data in a form so that it is easy to compare your results with expected or predicted values.

Discussion and Conclusion—This section summarizes key findings from the experiment and discusses points such as (1) to what extent the aims of the experiment were achieved, (2) important limitations of the experiment and any recommended improvements, and (3) the cause of any unexpected results.

Tips: Try to compare experimental results with any preliminary work or other published theoretical values.

Other sections that should be included in a report are the reference section and appendices. The **Reference** section is a list of references that are cited throughout the report. The **Appendices** are where additional content goes. A new appendix section should be created for each category of content. This next set of questions helps us consider guidelines for laboratory reports.

9.1.1. **Should you refer to figures, tables, or citations in an abstract? Explain.**
9.1.2. **Which aim statement below is more accurately written (Aim 1 or Aim 2)? Explain your answer.**

- **Aim 1: To conduct stress and strain measurements on metal specimens in a tensile machine.**
- **Aim 2: To measure stress and strain on metal specimens in a tensile machine.**

9.1.3. **For the following statements, identify which section it belongs in.**

- **The stress-strain graph showed that the copper sample experienced more plastic deformation than the steel sample, which was reflected in the higher percent elongation. This has been seen by previous researchers.**
- **The uniaxial tension test is the most common testing method for determining the mechanical properties of metals.**
- **The values of ultimate tensile stress for the steel and copper samples were 420 MPa and 210 MPa, respectively.**

- **An Applied Test Systems Corporation Testing Machine and Axial Extensometer were used to measure the force applied to and displacement of the samples.**
- **Uniaxial Tension Testing of Mechanical Properties of Metals.**

A **design report** is a document that reports the engineering design process and outcomes of a design task [5]. Its purpose is to communicate to a reader the understanding of the problem, evolution of the design, outcome of the design process, and next steps. Like a lab report, the design process can follow many strategies, but it typically begins with stating a problem or a customer's need. Then the main actions frequently include generating design concepts, analyzing the concepts, and narrowing and selecting a design concept to prototype, and then testing and redesigning a series prototypes to arrive at a final design. Different schools and eventually companies will have different formats required for write-up of the design report. The following list shows examples of various sections that could be included in a design report.

1. Abstract or Executive Summary
2. Introduction

    (a) Problem Background
    (b) Problem Statement
    (c) Constraints on the Solution

3. Design Approach

    (a) Generation of Candidate Concepts
    (b) Identification of components and hardware
    (c) Development of Preliminary Models
    (d) Generation and Optimization of Models
    (e) Selection of Design

4. Components and Hardware
5. Development of Preliminary Analytical Models
6. Advanced Modeling, System Simulation
7. Testing and Experimentation
8. Prototype Production and Manufacturing and Demonstration
9. Design Assessment

    (a) Requirement Achievements
    (b) Failure Analysis/Feasibility of Design
    (c) Product Enhancements/Future Recommendations
    (d) Life Cycle Analysis

10. Project Management Timeline (Gantt Chart)
11. Economic Analysis

    (a) Cost Considerations
    (b) Sales and Profit Considerations

12. Societal and Environmental Impact Analysis
13. Conclusions and Future Work
14. Project and Report Responsibilities
15. References

    (a) Engineering standards
    (b) Books
    (c) Technical journals
    (d) Patents

16. Appendices (As needed)

    (a) Detailed computations and computer-generated data
    (b) Manufacturers' specifications
    (c) Original laboratory data

This next sections let's us explore sections of a design report.

9.1.4. **Explain what a Gantt chart is and its importance.**

9.1.5. **Explain what a Life Cycle Analysis is and its importance in a design report.**

9.1.6. **Explain the importance of having a section titled "Societal and Environmental Impact Analysis."**

For both design and laboratory reports there are some common formatting guidelines and best writing practices to consider [2–5]:

Formatting Guidelines
- Technical documents typically are single spaced.
- Left justification or full justification is used in the document.
- One blank line is placed between paragraphs OR No blank link is placed between paragraphs and each paragraph is indented.
- Serif font such as Times New Roman is typically used. However, when documents are written for electronic media, a Sans Serif font such as Calibri or Arial is typically used. The font sizes are typically 11 or 12 points.
- One-inch margins around the text is standard.

Writing Practices
- When choosing words, consider the purpose of the document audience's level of knowledge regarding the topic.
- Write concisely in your own words and understanding. Reword any handouts that were given.
- Use past tense and passive voice when writing because past activities are being reported.
- Avoid personalizing the writing.
- The words that comprise acronyms and abbreviations need to be fully spelled out the first time they appear, with the shortened form appearing in parentheses immediately after the term. Then the abbreviation or acronym may then be used throughout the paper.

- Do not use vague language, give numerical precision when reporting or discussing results.
- Use analogies and metaphors to illustrate abstract or complicated ideas.
- Avoid using contractions because they are unprofessional and informal.
- Avoid generalizations because they are difficult to substantiate and are too broad to be supported in technical writing.
- Use gender neutral and generic terms when possible.
- Use objective terms and avoid including personal thoughts in technical writing.

Finally, inevitable there will be errors to fix so always remember to **proofread and edit a document several times before turning it in**. Finding lots of grammatical and formatting errors in any written document casts the author in a poor light. This section explores best practices for writing both laboratory and design reports.

9.1.7. **Explain the difference between Serif and Sans Serif font.**

9.1.8. **Explain what left justification and full justification of text are.**

9.1.9. **What is the following statement an example of? "Hydrogel material acts like Jell-O."**

9.1.10. **For the following statements, explain why it is not a best practice and rewrite the given statements using the writing best practices.**

- **We increased the speed of the motor.**
- **I feel this was the best method because it saved us an hour of time.**
- **The increase in stress was large.**
- **She measured the metal dog-bone specimen three times.**

9.2   Written Communication 2: Professional Writing

Engineers need to be proficient in professional writing as well as technical writing. **Professional writing** encompasses the variety of documentation that engineers write throughout their careers. While engineers may write any form of composition [6, 7], there are several common pieces of writing that engineers are likely to encounter, each with its own specific purpose or function.

Memos—short documents to either convey information, make announcements, or ask requests. For all types of memos, the information to be included is a *header* that includes the date the memo was written, who it is from, to the person or persons it is being sent, and subject of the memo. The body of the memo should be precise. Memos are typically two pages or less.

Proposal—a document that is the proposition of ideas for a solution to a problem or fulfillment of a need. It focuses on the motivation behind the project, background patent and literature search, and examines approaches to the solution. The format of proposals varies depending on the organization or sponsor of the project. Figure 9.1 relates progress reports to proposals and final reports.

	Proposal	Status Report	Final Report
General	Proposition of idea	Used to update manager (advisor) about status of project	Total package of product
Type of Information	• Focus on the motivation behind the project, • Patent and literature search • Approach to solution • Alternative Designs	• Less need to focus on motivation and background • Need to focus on Gantt chart and project tasks	• Contains most important information from proposals and status reports • All information from proposal and status reports is there, just reordered

Fig. 9.1 Differences between a proposal, status report, and final report

Progress Report—documents that communicate how much progress has been made and which project objectives have been achieved to date. Progress reports can be required as often as each week, each month, or per year, based on the type and timeline of the project. The format of progress reports varies depending on the organization or sponsor of the project. Figure 9.1 relates progress reports to proposals and final reports.

Final Report—document that communicates the culmination of engineering work. This could be an engineering design project, business project, or laboratory experiment. Formatting and guidelines for final reports are the same as what was taught in Sect. 9.1. Figure 9.1 relates progress reports to proposals and final reports.

Executive Summary—document that summarizes a longer report, proposal, or a group of related reports so that readers can rapidly become acquainted with a large body of material without having to read it all. Intended as an aid to managers and executives in decision-making. It is typically a few pages in format and can reference details in larger documents. Therefore, sometimes it comes in place of the abstract of a report.

Writing standards and guidelines for these types of documents are the same as for writing lab and design reports. And again, remember to proofread and edit a document several times before turning it in. This means that reports should be completed at least 2 days prior to when they are due so that ample time is allotted for proofreading and editing. This next activity helps us identify and compare the variety of types of engineering documents.

9.3.1. **Explain the similarities and differences between an abstract and an executive summary.**

9.3.2. **State which type of written format should be used to communicate the following information.**

- **Patent search on previous products**
- **Update workforce on quality standards for a product**
- **Alternative designs for a product**
- **Risk assessment and mitigation plan for a product design**

9.3   Visual Presentations

Presenting work is an essential part of an engineer's job. Once an engineer completes an analysis or a project, they need to be able to clearly present the results and outcomes. For displaying results, there are three main forms of presentation. Results can be displayed in a table, figure, or graph [7–12].

A **Table** presents numbers or text in list form in rows and columns. It is standard to use a table when raw data needs to be displayed or text needs to be organized in a precise manner. The components of a table include the title, column titles, and body of the table. Below are some best practices for structuring tables.

- Tables are centered on the page and come below the paragraph where they were first referenced.
- Table titles are numbered sequentially (Table 1) as they are referenced in the text, then followed by a clear description of what is in the table.
- Tables are numbered separately from figures.
- Table titles are labeled above the table.
- The column titles or row titles are labeled with a description of the data, including units of measurement.
- When reporting data in a table, watch reporting significant figures. Make sure to not over-report significant figures.

A **Figure** is a visual presentation of results. Figures can come in many forms such as chart, drawing, image, photo, map, or graph. Figures are used to display data and communicate processes or ideas simply and clearly. Below are some best practices for using figures.

- Figures should not replicate the information found in a table and vice versa.
- Figures are numbered separately from tables.
- Figures are centered on the page and come below the paragraph where they were first referenced.
- Figure titles are labeled below the figure.
- Figure titles are numbered sequentially as they are referenced in the text (Figure 1 or Fig. 1), then followed by a concise but comprehensive description of what is shown in the figure. The title can be used to draw attention to important features within the figure.

A **Graph** is a particular figure that shows quantitative relationships between variables or data. There are a variety of types of graphs that can be used depending on how to best display the data. Several common are listed below:

- **Bar charts**—used to show the relationship between independent and dependent variables, where the independent variables are discrete categories. In horizontal bar graphs, the dependent variable is labeled on the horizontal (x) axis, the independent variable is labeled on the vertical (y) axis.

- **Pie charts**—used to show the relationship of several parts to the whole. Use pie charts only when there are a limited number of mutually exclusive categories, and the sum of parts adds up to 100% of something. Make sure to include the percentages of each piece in the key.
- **Scatter plots**—used to display the relationship between two variables with points on a planar x–y graph. Each point represents one observation along the two axes. Also, they can be used to plot the mathematical relationship (linear or nonlinear) between the variables.
- **Line graphs**—used to display relationship between two variables by joining the points with a line. Used often when showing the rate of change between two points or comparing multiple dependent variables by plotting multiple lines on the same graph.

Always remember that the reader or viewer is going to need to understand and decipher the graph. Below is some useful advice regarding constructing graphs:

- Graphs are centered on the page and come below the paragraph where they were first referenced.
- Graph titles are numbered sequentially as they are referenced in the text. They are numbered as figures, not tables, then followed by a clear description of what is in the graph.
- Graphs titles are labeled below the graph.
- Label all axes and legends and remember to include units.
- Make graph axes and lines thick so that they can be seen.
- Include different symbols on the lines or textures of area instead of just having different colors distinguish between data, so that the graph can be understood if it is in black and white.

Finally, good practices for all visuals is to only include a visual if it adds value to the text. They are not just fillers for a paper or presentation. The visuals should be set apart from the text itself. Written text should not flow around a table, graph, or figure. The visuals always come after the first paragraph where the figure, table, or graph was referenced in. Finally, always cite the source of the data and/or figure at the end of the title if you did not create it yourself. This next activity examines applying best practices for the graphical presentation of data.

9.3.1. **Is every figure a graph? Explain.**
9.3.2. **For the graph in Fig. 9.2a, is a pie graph the best choice for representing the data? Explain your reasoning.**
9.3.3. **For the graph in Fig. 9.2b, what would be an appropriate figure description.**
9.3.4. **Based on best practices, describe what corrections should be made for the graph in Fig. 9.2c.**

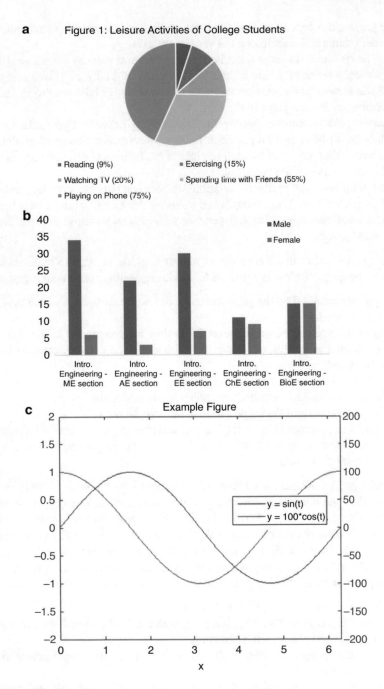

Fig. 9.2 Examples of different types of graphs (a) pie chart, (b) bar chart, and (c) line chart

Another form of communication that is prevalently used by engineers is visual aids. **Visual aids** are items of visual information used to enhance an oral presentation and include things like graphs, photographs, drawings, or video clips. Visual aids are chosen depending on their purpose. Engineers orally present the findings of their research or the results of their work at conferences and in business meetings, so having proficient skills in creating visual aids is a major benefit to an engineer. Two forms of visual aids engineers frequently employ are slide show presentations and poster presentations. **Slide show** presentations are a prearranged series of single pages (also known as slides) of information. The slides are projected on a screen or electronic display and are composed of text information, images, or charts. They are often used in business meetings and conferences. Currently, the industry standard to make slide shows is the software PowerPoint. However, recently alternative programs have emerged such as Prezi, Vyond, Zoho Show, Google Slides, Keynote, Haiku Deck, Canva, and Slidedog. A well-designed slide show can greatly help in delivering the key messages of a presentation to an audience. The following are a list of tips to help create effective slide presentations [13].

- **Words**—Keep words on the slide to a minimum. Do not put full sentences because it causes the audience to focus on the words instead of the speaker. Employ the *5/5/5 Rule*—No more than five words per line of text, five lines of text per slide, and five text-heavy slides in a row.
- **Pictures**—Should always have meaning and relevance on a slide. If a picture or an image is not being addressed or cited during a presentation, it should not appear on a slide.
- **Data**—Visual data as much as possible for ease of communication by creating graphs or figures for the numbers being presented.
- **Font**—Text should be pleasant to look at and easy to read. Simple fonts include San Serif fonts such as Calibri, Arial, or Verdana. Also make sure the font is sized aptly so that it is easy to read, but not overwhelming. Suggested sizes for headings are typically minimum of around 20 pt, and for the body a minimum of 16–18 pt is recommended.
- **Color Scheme**—Including a color scheme livens up a presentation but define a few colors as key contributors. When many different colors are used without order, it makes the presentation look juvenile and amateurish.
- **Background**—It is important that the text on each slide be seen, and therefore it is recommended to leave the background of a slide plain and white. If a colored background or picture is used make sure the text is readable by either boarding it, changing text color, or shadowing it.
- **Animation**—many software packages will have the ability for there to be animations on a slide for things like emphasizing text or transitioning between images. Do not overload a presentation with animations. Use the simple animations in order to have the audience follow along with your presentation.
- **Finally Important**—Do not read your presentations straight from the slides. Slides are a reference for the presenter but mainly a visual for the audience, so talk along with the slides while the audience views them.

Poster presentations are often used in academic conferences. A large one-page display is created to summarize information concisely and attractively. A poster typically contains a combination of text, graphs, pictures, and diagrams to explain the story of the work. At research conferences, the researcher typically stands by their poster display while other conference participants walk around and view the presentation and interact with the author. Typically, posters are made by using any of the previously mentioned slide show software packages, using one slide, and formatting so it can be printed at a large scale. Since a poster is essentially a large slide, the font, color scheme and background guidelines of a slide presentation apply. The mark of truly good posters is that it can be viewed and understood when the presenter is not there to talk about the work. Below is a list of the main sections included in a poster and some best practices when creating them [14].

Title—Should succinctly convey the issue, the general experimental approach, or the system being discussed. Try to make the title catchy so that participants will be interested in reading the poster.

Introduction/Background—Explain the issue or question while using the absolute minimum of background information and definitions. If possible, convey with chart, photograph, or illustration that communicates some aspect of the work.

Solution/Design—Describe the overall final solution or design produced. Include chart, photograph, or CAD drawing of the final solution or design. Make sure to label and describe important components.

Experiments/Results—Describe experimental setup and important analysis performed. This also should be conveyed with images or flowcharts as much as possible.

Conclusions—Summarize the major results and findings of the work, the relevance of the findings to other published work and future directions for the work.

Literature Cited—Previous works by others (journal articles, books, etc.) where intellectual content was taken and stated in the work.

Acknowledgments—Include individuals for specific contributions such as equipment donation, technical guidance, and laboratory assistance. Make sure to write full names. Also, include who has provided funding for the work. This section can also include disclosures for any conflicts of interest and conflicts of commitment.

When done correctly both slide shows and poster presentations can grab viewers' attention and immensely improve a presentation. This next activity explores slide and poster presentations.

9.3.5. **For each pair of slides in Fig. 9.3, choose which slide is better formatted and explain why.**

9.3.6. **Explain what good and what improvements can be made for the posters in Fig. 9.4?**

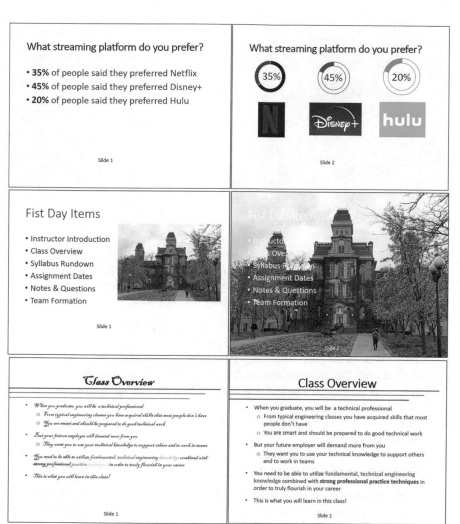

Fig. 9.3 Examples of slides from oral presentations

9.4   Oral Presentations

As important as it is for an engineer's written work to be understood, it is equally important that an engineer's spoken words be easily understood. Having good oral presentation skills is relevant for both academia and industry. It is critical for an engineer to have good oral presentations skills so that the listeners understand the work performed. If the audience does not understand the results being conveyed, it renders the work essentially insignificant. While some people are more comfortable speaking to an audience than others, there are strategies and methods that can be followed to give the best presentation possible.

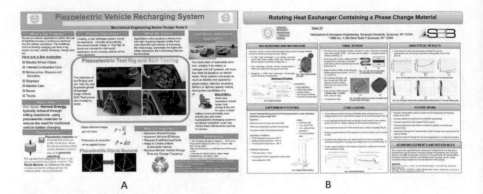

A   B

Fig. 9.4 Examples of poster presentations

The first step in creating a good oral presentation is **preplanning** [15, 16]. In this step, the presenter thinks about how to tailor presentation based on audience, time, resources, and desired outcomes. A good practice in preplanning is to answer four "W" questions and one "H" question. Namely the following:

- **Who**—Who is the audience—the age range, demographics, professional positions, and knowledge on the topic to be presented.
- **What**—What is the purpose of the presentation—what needs to be accomplished and what does the audience need to take away from the presentation.
- **Where**—Where is the presentation taking place—room size, room layout, and equipment available.
- **When**—When does the presentation take place—time during the day and order of other presenters.
- **How**—How long should it take—is there a time limit on the presentation, when will I lose the audience's attention.

The second step is **preparing** the visual and oral elements of the presentation. For the guidelines on the visual elements refer to the previous sections of this chapter. For the oral elements, there are two components to tackle

1. **Layout**—Each presentation will be different, but there are core elements to address in each presentation, namely
   - *Introduction*—states the background and purpose for the talk
   - *Body*—the main information for the talk
   - *Conclusion*—summarize major points, future work, show appreciation for the audience's attention, and solicit questions from the audience

2. **Wording**—the arrangement and language in a presentation is vital in keeping an audience informed and engaged. When working on the wording for a presentation, one best practice is the 4-S formula [16]. This practice means try to keep the words and sentences

- *Short*—keep to the main subjects and avoid going off on tangents
- *Simple*—lengthy words and phrases may confuse and bore people
- *Strong*—using action verbs and an active voice
- *Sincere*—respect and connect with the audience

The third and final step is **practice**! It is a good habit to script out the main points that are addressed on each slide so that lots of practice can happen prior to the presentation. Practice in front of a mirror, then in front of friends or family. Also, time the presentation to make sure it will fall within any given time constraints. Although writing or typing a script for practice is encouraged, a presenter should never bring a written script or note cards to the actual presentation. It makes them look unprepared.

When a presenter is well prepared, gives a considerate, organized, and precise presentation the audience will greatly enjoy it and learn a lot from it. This next set of questions looks at best practices for oral presentations.

9.4.1. **From the list of sentences given, label whether the sentence is written in active or passive voice.**

- **Auggie did his math homework**
- **The bridge was built by engineers**
- **Sarah types up the statics homework**
- **The stress in the beam was measured by the lab students**
- **The report has already been discussed by the team**
- **The professor writes on the whiteboard**

9.4.2. **For the sentences in the previous question that were passive voice, rewrite them using active voice.**

9.4.3. **Which sentence would you rather use in a presentation A or B? Explain your answer.**

> **A: While promenading around the provincial reservoir, my auditory nerve perceived a plethora of wildfowl honking vociferously.**
> **B: I heard a lot of ducks quacking loudly when I was walking by the lake.**

9.4.4. **From the list of sentences given, label which section of a presentation it should appear in, introduction, body, or conclusion.**

- **Poly(vinyl alcohol) (PVA) hydrogels are a class of biomimetic polymers that swell in the presence of fluids, yet are insoluble due to their cross-linked structure.**
- **In the future we plan to combine both approaches to create a hydrogel with exceptional, self-sustaining low-friction surface properties.**
- **To chemically assess our gels, we used Fourier Transform Infrared Spectroscopy.**
- **The first objective of this study was to develop a method for the synthesis of zwitterionic polymer brush functionalized hydrogels.**

- **Mechanical Stiffness was evaluated from confined compression experiments.**
- **I appreciate the time you took to listen to my talk today, and now I would like to invite any questions people have.**

9.5   Team Communication

Communication is an essential part of life and humans communicate in many ways. So, while giving formal presentations is part of an engineer's career, there is also lots of informal communication that occurs. Particularly when interacting as part of a team. There are many conceptual models to describe the communication process [17–26], but there are four common fundamental aspects to communication as shown in Fig. 9.5. The **Sender** creates and transmits the message or messages. The **Receiver** is the person (or group) who receives and responds to the message. The **Message** contains both the sender's information and the receiver's interpretation of the message. Finally, **Feedback** is the way the receiver acknowledges or responds to the message. In oral communication, it is important to realize that a message is not just what a person says, but the meaning of the message, and meaning comes in

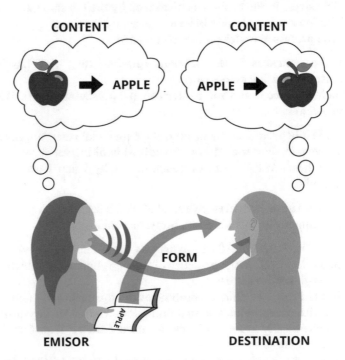

Fig. 9.5   A general model of the communication process

Table 9.1 Common bridges and barriers

Communication bridges	Communication barriers
Listening well	Not listening
Encouraging verbal or nonverbal response	Yelling
Solid eye contact	Getting angry
Expressing Empathy	Lying
Repeating the message given back to the sender	Sulking
Asking clarifying questions	Acting sarcastic
Remaining relaxed	Name-calling
	Rolling eyes
	Interrupting
	Acting Threatening/Accusing/Blaming

two forms: (1) that is intended by the sender and (2) what is interpreted by the receiver. If those two meanings do not align then there is *miscommunication*. That makes feedback important for clarifying any questions between the sender and receiver.

There are two elements that can affect communication between people. They are typically known as Bridges and Barriers [27]. **Communication Bridges** are positive things that people say or do to help open and ensure communication continues. **Communication Barriers** and negative things that people say or do that prevent listening and understanding. There are elements of bridges and barriers when everyone communicates. The balance between the two elements is what causes communication goes to go well or poorly. Table 9.1 lists common bridges and barriers.

To foster positive communication, particularly when a delicate or divisive subject is being discussed it is important to express feelings honestly and clearly without threatening or insulting people. A key practice is to **begin the communication with "I" (I wish, I feel) rather than "you" (you always)**. This way, the sender is taking ownership of the situation and the receiver is more likely to respond positively. Another area where students sometimes struggle is effectively *communicating directions*. To communicate directions in a nonthreatening way (1) give very detailed directions (step-by-step) and (2) constantly ask for feedback about the directions.

Poor communication can cause many problems in a person's personal and professional life, so it is good to build strong communication strategies. This next activity helps learn and practice good communication basics.

9.5.1. **List as many ways to communicate as you can think of.**
9.5.2. **For the following negative communications, write a positive alternative that still conveys the same message.**

   - **"You never return my calls."**
   - **"That is a stupid idea."**

- "Nobody cares how hard I work."
- "You always ignore my design ideas."
- "Do not yell at me."

9.5.3. List any communication bridges you can think of that are not listed in Table 9.1.

9.5.4. List any communication barriers you can think of that are not listed in Table 9.1.

9.5.5. Give an example of a conversation you have had recently, identify the bridges and the barriers within the conversation and explain the results of the communication.

9.5.6. For the following communications, state the attribute of the statement.

- Why do you always have to be late for every meeting?
- You could have been a little later, then you had missed the entire meeting!
- You are such a dumb loser for being late.
- We are not making meeting plans with you in mind ever again.
- Now we will never complete this report and you have ruined the meeting.

One of the most important traits an engineer can learn is the ability to work as part of a team. Teamwork is defined as work done by several individuals cooperating to accomplish a common goal or solve a problem. Working in a team has many benefits including diversity of ideas and skills. However, there is also the potential for conflict. When managed properly conflict can be productive. To work together successfully, group members must foster and establish a sense of cohesion. There are processes and strategies that can be employed so teams work together effectively. There is lots of information about successful teamwork strategies [28], but below discusses four common important tasks a team must establish and strategies to be successful.

1. Purpose and Roles

Every member of the team needs to agree on *what needs to be done* and by *whom*. Then the team plans the tasks needed to accomplish the goal and each member takes responsibility to complete certain tasks. Once the purpose and tasks are established, members can hold each other accountable for completing their tasks. A common strategy for this is to develop a project timeline. The **project timeline** includes the overall goal and the breakdown of tasks for the project. It can also include a responsibility matrix where the project tasks are listed along with the team member(s) responsible for completing the task and the due date of each task.

2. Process

Teams need to develop strategies central to how their group operates. This includes how group decision-making is facilitated, how productively is tracked, and how conflict is addressed. They also need to establish processes for team communication and meeting organization. One standard strategy is employed in the creation of a team contract. A **Team Contract** is a document prepared by each team prior to starting work on group projects. The members develop their own "rules of engagement." It provides an opportunity for groups to specify preferred methods of communication, action plans, meeting schedules, goals, and consequences of actions (or inactions) of group members. An example of a team contract is shown in Fig. 9.6.

Team Contract

Project Title _____

	Team Member Name:	Email	Other Contact Info:
1			
2			
3			
4			
5			

THINGS TO NOTE:
- You cannot change any aspect of the team contract, particularly the "Procedures" section, once the signed contract has been turned in to me. Therefore, be VERY SPECIFIC on the Procedures section and spend some time thinking about this section. The only way a change to your Team Contract can be made is if ALL members of your team sign and date a written amendment to your initial agreement.

A. Team Structure

1. Leadership structure (individual, individual with rotating leadership, shared).

2. Decision-making policy (by consensus? by majority vote?).

3. Who is the team Recorder/Documentation Manager/Maintainer of all required turn-ins?

4. Day, time, and place for regular team meetings.

5. Usual method of communication (e.g., e-mail, cell phone, wired phone, in person).

Fig. 9.6 Example of a team contract

B. Team Procedures
1. Method for setting and following meeting agendas (Who will set each agenda? When? How will team members be notified/reminded? Who will be responsible for the team following the agenda during a team meeting?)

2. Method of record keeping (Who will be responsible for recording and disseminating minutes? How and when will the minutes be disseminated? Where will all agendas and minutes be stored?)

3. Procedures in the absence of a team member: (will the team meet with one member absent, or must all members be present?)

C. Team Participation
1. Strategies to ensure cooperation and equal distribution of tasks.

2. Strategies for encouraging/including ideas from all team members (team maintenance).

3. Strategies for keeping on task (task maintenance).

D. Personal Accountability
1. Expected individual attendance, punctuality, and participation at all team meetings

2. Expected level of responsibility for fulfilling team assignments, timelines, and deadlines.

3. Expected level of communication with other team members.

4. Expected level of commitment to team decisions and tasks.

E. Consequences of Breach of Contract
What procedures and penalties do you wish to implement in the case of "slackers" or team members who deviate from your Team Contract? You basically have two options here:
1.) No Peer Evaluation or
2.) Peer Evaluation.
Either option is acceptable, but the instructors strongly advise you implement Peer Evaluation so that you have a mechanism in place if problems arise.

   1. No Peer Evaluation: Your group may decide that you do not want any form of peer evaluation and that you will trust each other to pull their weight throughout the entire semester. If that is your decision, in this section of your contract, simply specify the following statement: *All group members will receive the same grade*

Fig. 9.6 (continued)

on every aspect of the project regardless of their contribution and regardless of any problems that may arise throughout the semester.

2. <u>Peer Evaluation</u>: If you would like some form of peer evaluation, specify the following statement: *Our team will conduct peer evaluation as part of this project.*

Note: If students implement a percentage system. The percentages need to be agreed upon and signed by all team members prior to handing in an assignment.

1. Policies for handling infractions of any of the obligations of this team contract.

2. Policies for handling persistent infractions.

F. Certification by team members
In appending your signatures below, you are stating that:
   a) You participated in formulating the standards, roles, and procedures of this contract;
   b) You have agreed to abide by these terms and conditions of this contract;
   c) You understand that you will be subject to the consequences specified above and may be subject to reduction in overall course grade in the event that you do not fulfill the terms of this contract.

Printed Name:			
Signature		Date:	

Printed Name:			
Signature		Date:	

Printed Name:			
Signature		Date:	

Printed Name:			
Signature		Date:	

Fig. 9.6 (continued)

Another part of the process is that group members know how to run a meeting. Meetings are key events during group work because when run correctly they can help manage group dynamics, promote productivity and foster comradery. There are several techniques for running an effective meeting [29], but below are some best practices.

Before the Meeting

- Plan the meeting carefully: who, what, when, where, why, and how many
- Prepare and send out an agenda, identifying issues to be discussed and a timeline for discussion

During the Meeting

- Start on time
- Clearly define agenda items, time limits, and member roles
- Review action items from previous meeting
- Focus on one issue at a time for the allotted time only
- Document the meeting either through written notes or recordings

After the Meeting

- Record final decisions or actions that need to be taken
- Assign tasks to group members until the next meeting
- Set deadlines for the tasks
- Set the date and place of the next meeting and develop a preliminary agenda
- Close the meeting positively and clean up the meeting room
- Prepare a meeting minutes memo, distribute to members, and others for feedback and ratification

3. Ways to Decide

Part of the process of a successful team is establishing rules for how to make decisions and deal with conflict. Being able to come to an equitable decision as efficiently as possible is important for the functioning of a team because the overall performance of a team involves accounting for the needs and opinions of everyone. There are a variety of ways to make decisions as a group but listed below are some suggested methods for how to reach a decision. These can also be effective methods for dealing with conflict that arises within a team.

- Decision by majority—the team holds a vote, and the majority vote wins.
- Decision by unanimity—all members must agree that the decision is the best one.
- Decision by ranking—members individually write down the ideas they like best, then rank each idea from highest to lowest. The votes for ranking are tallied and the idea with the highest total is selected.
- Combining ideas—the team searches for possibilities of implementing both or combining ideas into one solution.

4. Accountability and Feedback Mechanisms

Once the measures and rules for productivity are formed, mechanisms need to be established for receiving feedback from the team and consequences for failure to meet the required task. Some activities for accountability and feedback include team contracts (which were previously mentioned), utilizing task management tools and logs, or group accountability scoring tools. One example of a group accountability tool is the Comprehensive Assessment or Team-Member Effectiveness (CATME) tool [30]. CATME is an online program that is used in team formation and as a feedback tool for teams. Sometimes it is prescribed as part of a class, but students can use it for free as a team tool. Ultimately it is up to each individual team how to track and measure productivity.

Good teamwork is important in engineering because the world is more connected than it has ever been and the problems that need solving are complex. This next activity explores and reflects on effective teamwork practices.

9.5.7. **Explain a good experience you had working in a team.**

9.5.8. **Explain a bad experience you had working in a team.**

9.5.9. **For the previous two answers, explain the similarities and differences between the two experiences.**

9.5.10. **Go to the website https://info.catme.org/and list the five different types of contributions to a team that CATME rates team participation on.**

9.5.11. **If a teammate does not complete tasks as required by the team, discuss strategies you would employ to address the matter.**

9.5.12. **Answer the following questions with your preference when working on a team.**

- **How much do you feel you can rely on your group members to complete a required task?**
- **How do you ensure that all voices on the team are heard?**
- **How do you make sure that group members feel supported, encouraged, and appreciated for their work?**
- **How do you encourage the team to stay accountable to the tasks they have been assigned?**
- **What do you do if a group member is unhappy or uncomfortable with a decision made by the group?**

End of Chapter Problems

IBL Questions

IBL1: For the following slide presentations, based on best practices answer the following questions:

1. Describe the positive attributes of the presentation.
2. Explain strategies to improve the presentation.

Integrated Use of Programing in Machine Design Course

Amy Bloom, Ph.D., Todd Shaffer, III, Ph.D.,

Mechanical & Aerospace Engineering Department

Syracuse University

2017 ASEE Annual Conference and Exposition
June 26, 2017
Columbus, Ohio

Overview

- Background
- Objectives and Goals
- Design Project Descriptions
- Instructor Materials
- Discussion and Student Outcomes
- Future Projects and Conclusions

Background

- Machine design is a required course for all nationally accredited mechanical engineering undergraduate Bachelors of Science degrees.
- The course typically occurs during student's junior year, and serves as the key link between the engineering mechanics course sequence consisting of statics, dynamics, and strength of materials with the capstone project course for senior students.
- Over the years, the machine design course has been improved with efforts to introduce
  - Project-based learning [Steuber, J.G., 2011]
  - Hands-on machine design laboratories [Monterrubio and Sirinterlikci, 2015]
  - Finite element analysis projects [Liu and Brown, 2008]
- However, in engineering practice, only some of an engineer's time is spent executing analyses that have been pre-packaged. More often than not, the engineers find themselves in the position of performing an analysis that is different from those done previously.
- Hence, there is a great need for students to learn how to comprehensively program software to solve machine design problems because it allows them to simultaneously consider multiple design considerations.
- While there are current practices utilizing computational tools in a machine design course, most of the student interaction with the program is where the students only needs to input initial design parameters, and all other design calculations are performed within a program [Kim and Rezaei, 2008]
- Therefore, it would be beneficial to teach students how to implement their theoretical understanding into an analysis and by doing so, the fundamental theories and concepts are not marginalized.

In this Class: First Half = Design Analysis

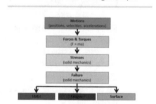

In This Class: Second Half = Machine Synthesis

Machine Elements = Consider only the mechanical parts of machine's and their analysis

Mechanisms (linkages) = subsystems of a machine

- Types for the semester
  - Bar Linkages (motions)
    - Can be moving or stationary
  - Shafts, Keys, Couplings
  - Bearings
  - Gears
  - Springs
  - Screws and Fasteners

Objectives and Goals

Objective: The objective of introducing the MATLAB projects was to encourage young mechanical engineers to enhance their programming skills, because modern day engineers need software programming skills to be successful and have a professional edge in industry.

Goal: The goal of the projects was to provide a platform for improved understanding of machine design and emphasize the importance of programming in mechanical engineering, specifically
(1) the fast speed at which complex problems can be analyzed
(2) the ability it provides students to be able to develop creative solutions

Project #1

Introduction to Design and Design Factc

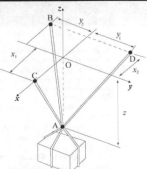

- Scenario: two crates, each with different masses (1,000 kg, 2,000 kg), and different size diameter of cables (varying from 10 mm to 30 mm)
- Students asked to investigate how various combinations of weight and diameter effect the distance from the ceiling that the boxes could be supported. D
- Design the diameter of cable that will be able to support the to ceiling as possible but not violate a safety factor of 3.

Purpose of this design project was to give students an opportunity to compute the distribution of safety factor over a changing design parameter, and make engineering decisions based on the information obtained

- Students use concepts from previous classes (statics, solid mechanics and computational methods)

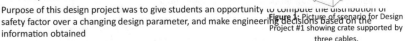

Figure 1: Picture of scenario for Design Project #1 showing crate supported by three cables.

Project #2

Kinematics of a Linkage System

- The analysis requires students to code the 6-bar linkage mechanism to evaluate the position, velocity and acceleration of the various joints.

- Then, the students were required to prepare a formal report documenting their analysis, including a discussion on the correctness of their results and comments pertaining to the function of the mechanism.

Purpose: Give students an opportunity to compute the displacements, velocities, and accel of a linkage system and use modern engineering tools to reduce analysis time

- Students used concepts learned in class (fundamental theory of linkage kinematics)

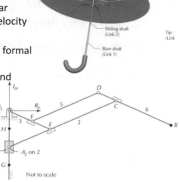

Figure: Picture of scenario for Design Project 2 and 3 showing (top) an umbrella device (b) the mechanism used in a folding umbrella expressed as links with joints A thru H and O_3.
[Cleghorn and Dechev, 2015]

Instructor Materials

- The projects can be performed with several computational softwares
  - Maple (Waterloo Maple Inc, Waterloo, Ontario)
  - Mathematica (Wolfram Research, Champaign, IL)
  - TK Solver (Universal Technical Systems, Inc, Loves Park, IL)
- To date, the computational programming has been performed, primarily in the software MATLAB (MathWorks, Natick, MA)
- Students should be nominally separated into teams of no more than 4 (2-3 students per group is ideal)

Student Outcomes

• *Survey given to seniors students who took machines design in spring 2016*

MEE332 EVALUATION SHEET
MATLAB Design Projects

Please indicate the semester you took MEE332: _____
Please answer the following questions by circling the number (1-5) that best describes your thoughts.

1. How would you describe the pace and schedule of the MatLab projects?
(1 = much too slow/lagged behind material taught in class, 3 = just right/stayed with material taught in class, 5 = much too fast/had to accomplish them before we were taught class material)

| 1 | 2 | 3 | 4 | 5 |

2. How would you describe the level of difficulty of the MatLab projects?
(1 = much too easy, 3 = just right, 5 = much too hard)

| 1 | 2 | 3 | 4 | 5 |

3. How would you rate the intellectual stimulation you received from the MatLab projects?
(1 = not enough, 3 = just right, 5 = much too much)

| 1 | 2 | 3 | 4 | 5 |

4. Please rate how helpful the MatLab project were at showing you the fast speed at which complex problems can be analyzed.
(1 = not at all, 3 = somewhat, 5 = extremely)

| 1 | 2 | 3 | 4 | 5 |

5. Please rate how well the MatLab projects provided you with the ability to be able to develop creative solutions for design problems.
(1 = not at all, 3 = somewhat, 5 = extremely)

| 1 | 2 | 3 | 4 | 5 |

6. Please rate how much the MatLab projects motivated you to increase development of your programming abilities.
(1 = not at all, 3 = somewhat, 5 = extremely)

| 1 | 2 | 3 | 4 | 5 |

Student Outcomes

• Question 1: how would you describe the pace and schedule of the projects?
  Question 2: How would you describe the level of difficulty of the projects?
• Question 3: How would you rate the intellectual stimulation you received from the projects?
• Question 4: How helpful the projects were at showing you the fast speed at which complex problems can be analyzed?
• Question 5: how well the projects provided you with ability to develop creative solutions
• Question 6: How much the projects motivated you to increase development of your programming abilities.
(1 = worst ranking, 3 = middle 5 = best)

Table 1: Percentage (%) of student at each ranking level for each question.

Rankings*	Question 1	Question 2	Question 3	Question 4	Question 5	Question 6
1	0%	3%	0%	2%	7%	8%
2	3%	5%	15%	15%	25%	19%
3	63%	42%	64%	42%	41%	32%
4	25%	44%	15%	34%	24%	27%
5	8%	5%	5%	7%	3%	14%

Student Examples and Outcomes

SYRACUSE UNIVERSITY ENGINEERING & COMPUTER SCIENCE

Conclusion

- The students appreciated the coding opportunities that the projects provided.
- Working together in small groups encouraged the students to discuss the codes and results.
- From the results of the questionnaire, it was evident that an improved understanding of machine design was successfully tied into design-based projects, while simultaneously emphasizing the importance of programing in mechanical engineering.

Future Projects

- It would be beneficial to add one more programming Design Project which focuses on on uncertainty and ambiguity in the design and analyzing multiple design parameters
- One idea for this project is to investigate the dimensional and material design of a simple machine such as a pair of ice tongs.
- Students would be given initial geometry of the system along with the weight and dimensions of blocks of ice.
- Students would be asked to design the tongs for cases of static failure and fatigue failure, investigating different cross sectional areas and multiple material properties, including ductile and brittle materials.
- Students would use concepts they have learned in classes such as curved beam mechanics, principle stress analysis, von Mises and Coulomb-Mohr effective stresses, as well as fatigue failure theories.

A Model for a Faculty Development Course Redesign Summer Working Group

Amy Bloom, Elizabeth Dougherty and Jessica Michalenko

Engineering &
Computer Science
Syracuse University

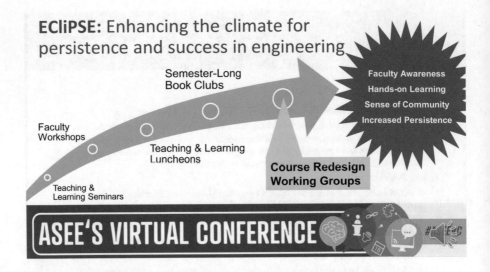

Requirements of working group participation

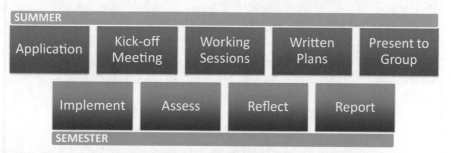

Typical 2-hour working session

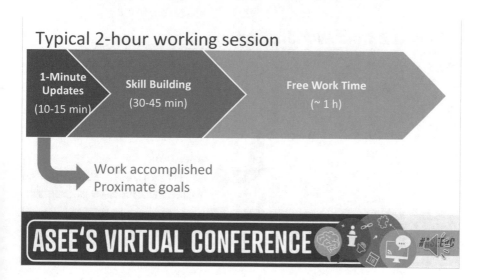

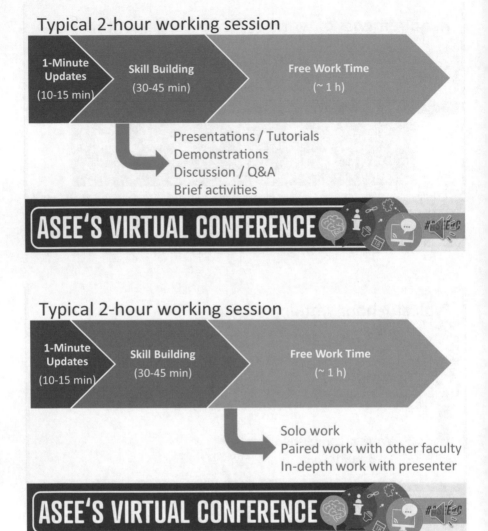

Project deliverables required in the summer and after implementation semester

JUNE
- Draft outcomes
- Prelim. assessment plan

AUGUST
- Presentation

JULY
- Student learning outcomes
- Implementation plan
- Assessment plan
- Example work

JAN / JUN
- Project Assessment
- Student Assessment
- Reflection

ASEE'S VIRTUAL CONFERENCE

Scope and type of redesign projects in 2019

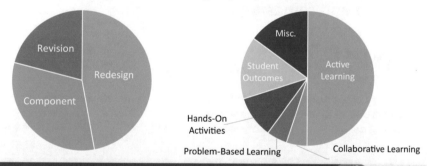

Revision
Redesign
Component

Misc.
Student Outcomes
Active Learning
Hands-On Activities
Problem-Based Learning
Collaborative Learning

ASEE'S VIRTUAL CONFERENCE

Scheduling for 20 faculty over 9 weeks is difficult, even in the summer

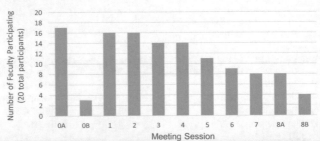

Most faculty completed all requirements in the summer and semester

95%

of original participants
completed all summer
obligations

90%

implemented and
assessed the next
semester(s)

80% *

reported on student
outcome assessments
*COVID-19 disrupted some
Spring 2020 plans

Faculty feedback overwhelmingly positive

Most valued program aspects:
- "Skill Building" time
- Accountability
- Learning from peers

"It was also nice to see what other people's projects were. It gave me ideas for some of my other classes."

"...each speaker did a very good job of speaking to the practical application and implementation of each topic. No one spent a long time...trying to convince us to use the method..."

"...having the college demonstrate an interest in investing in us as teachers and our courses is sometimes all it takes to prompt us to go above and beyond the norm."

Plans for future summer offerings

Deployment of a common student survey to participating classes

Reduction of total stipend

Participant input on Skill-Building session topics

Engineering &
Computer Science
Syracuse University

NSF Graduate 10K+ Award No. 1317540

Meeting the Graduate 10K+ Challenge: Enhancing the Climate for Persistence and Success in Engineering (ECliPSE)

IBL2: For the following poster presentations, based on best practices answer the following questions:

1. Describe the positive attributes of the presentation.
2. Explain strategies to improve the presentation.

Physical, Mechanical and Tribological Characterization of Boundary Lubricant Enhanced Hydrogel for Biomaterials Applications

Allen O. Osaheni, Michelle M. Blum
Mechanical and Aerospace Engineering Department, Syracuse University
College of Engineering & Computer Science

INTRODUCTION

- Hydrogels are a class of synthetic biphasic insoluble polymers that are used in a variety of biomaterial applications such as artificial tissue implants and scaffolds. [1]
- One major limitation to their performance has been that their surface and friction properties are incompatible with natural biological surfaces, leading constructs to wear, fail, and damage opposing contact surfaces. [2]
- This incompatibility issue is being addressed by blending a hydrogel polymer with a zwitterionic polymer, which acts as a boundary lubricant.

- **Objective: Conduct foundational studies that evaluate the physical, mechanical and tribological properties of the novel biomaterial.**

MATERIAL SYNTHESIS

- Biocompatible hydrogel polymer is blended with a zwitterionic polymer, which acts as a boundary lubricant. **(Fig. 1)**

Figure 1: (a) PVA & pMEDSAH components of system. (b) gel fabrication procedure where PVA & pMEDSAH dissolved in DI-water at 90°C and solution is physically crosslinked through a series of freeze thaw cycles.

CHEMICAL CHARACTERIZATION

- ATR displayed presence of the ester peak characteristic of pMEDSAH 1735 cm⁻¹ in the transmission spectra of PVA showed that the functionalization was successful. **(Fig. 2)**

Figure 2: ATR spectra of neat PVA and PVA-pMEDSAH blend.

PHYSICAL & MECHANICAL CHARACTERIZATION

- Blending pMEDSAH did not significantly influence contact angle or water content. **(Fig. 3a)**
- Water content was comparable to reported values for natural tissue, such as articular cartilage. **(Fig. 3a) [3]**
- Friction coefficient (COF) is reduced while maintaining a compressive modulus above 1MPa. **(Fig. 3b)**

Figure 3: (a) Water content and contact angle and (b) modulus of increasing PVA-pMEDSAH blend

TRIBOLOGICAL CHARACTERIZATION

- Decreased COF ranged from 80–86%. 5% pMEDSAH saturation point. **(Fig. 4a)**
- Average hydrogel surface roughness (R_a) ranged from 0.36–0.66 μm.
- COF reduction maintained for ~6 hrs with little as 3% pMEDSAH. **(Fig. 4b)**
- Film thickness estimated as low as 13 nm [4], thus pMEDSAH molecules shown to influence the friction characteristics in boundary lubrication (λ~0.005).

Figure 4: (a) Average COF of increasing PVA-pMEDSAH blend (n = 5). (b) Representative COF vs time for the neat and blended material.

FUTURE WORK

- Investigate feasibility of a self-replenishing boundary lubricant hydrogel. **(Fig. 5)**

Figure 5: (a) A healthy surface gets (b) damaged and removes the functionalized brush layer, the embedded lubricant gets drawn into the site due to the change of surface entropy. (c) inter-chain association causes chains to fuse across the damage site.

REFERENCES

1. Peppas N.A., 1987, CRC Press.
2. Blum and Ovaert, 2013, Wear, 301 – 209.
3. Baker et al., 2012, JRME Part B, 1453 – 457.
4. Gratin A.N., 1949, TUIROTMMS, 115-166.

Syracuse University presents Howell Venture's Featherlite 2.0

Team 11: Ben Rohde, David Brisson, Ryan Kearns, Sam DenDanto, and Jeffrey Chiu

Special Thanks to: Cody Howell, Michelle Blum, Frederick Carranti, Jürgen Babirad, Dane Millar, Ben Jones

INTRODUCTION

There are more than 40 million Americans who have some form of a disability, and over half of those people suffer from a leg mobility issues. An automobile hand control device enables people with leg mobility issues to drive by only using their hands. A hand control device is defined as any device that's used in an automobile to control the vehicles braking and throttling. There are two main types of hand controls that are used in the automotive industry. The first is a mechanical hand control device, which has been used for decades. The second, and more contemporary, device is an electrical hand control device. This project focuses on the electrical variety. Figure 1 is a blow out of what a fully assembled hand control would look like and figures 2 and 3 were the focus of what Team 11 redesigned in conjunction with Howell Ventures. In this project we used several different software resources at Syracuse University including Solidworks and Ansys.

OBJECTIVES

- Designing the springs to return to its neutral position from any accelerator position in less than one second
- Reduce the number of springs in the system from three to two by moving the springs back to the main shaft that connects to the sensors
- Effective EMI shielding to protect the system from any magnetic interference
- Adherence to all federal automobile regulations pertinent to this project
- Limited to 20 dollars of manufacturing costs

FEATHERLITE 2.0

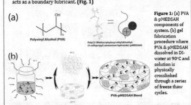

Figure 2

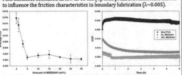

Figure 3

Part	Cost	Description
1-Music wire torsion spring	$4.00	OD 1.2", No Coils 6.25
2-Music wire torsion spring	$2.00	OD 0.2", No Coils 4.25
3-Hall effect sensor	$3.00	Dim 0.2"x0.23"
4-Alloy magnet	$2.00	Dim 3/8"x3/4"x1/12"
5-Gas plate and Bracket	$6.00	Sheet Metal 3/16"
6-POM enclosure & HEMMP	$1.00	Polyoxymethylene injection molded
7-Set Screw	$0.50	Dim 1/8"x1"

ANALYSIS

Figure 4
This figure shows a stress analysis using finite element analysis using Ansys software. As can be seen the maximum stresses on the HEMMP is around 540 psi, which is much lower than POM's ultimate stress.

Figure 5
This figure shows the safety factor again using finite element analysis using Ansys software. Under normal working conditions the HEMMP has a safety factor of at least 15 which should make this piece considerable more robust than required.

HOW IT WORKS

The user inputs vertical displacement into the hand control which is then transferred as an axial displacement to the HEMMP. Attached to the HEMMP is a small magnet which revolves around a Hall Effect Sensor. Since the magnet is rotating this creates a change in magnetic flux around the area of the magnet. This change in magnetic field alters the voltage of the current inside the Hall Effect Sensor. This voltage is then sent to the automobile's ECU which then correlates said voltage to an acceleration.

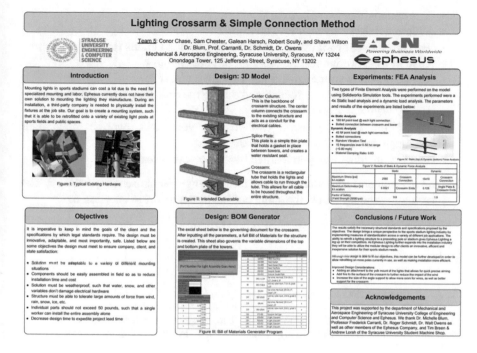

Figure I: Typical Existing Hardware

Figure II: Intended Deliverable

Figure III: Bill of Materials Generator Program

Practice Problems

1. Using standard practices and good technique, write a lab report for the IBL1 activity in Chap. 7.
2. Using standard practices and good techniques, create a poster for the IBL1 activity in Chap. 7.
3. Using standard practices and good technique, write a design report for practice problem 2 in Chap. 7.
4. Using standard practices and good technique, create a poster for practice problem 2 in Chap. 7.
5. Write a memo to your professor discussing their assignment or homework policy. State whether you agree with it and the reasons why. Or propose a new policy and explain why you think it is superior to the current policy.
6. Submit a six-page report about a famous engineer. Your report should include but is not limited to the following:

    • The first, middle, and last name of your engineer.
    • A picture of your engineer.
    • Years living/lived.
    • Their field of study.

- The country they were born in
- Where did they live/what was their job?
- Explain (in your own words) what the engineer is famous for.
- Give at least two interesting facts about your engineer.
- Give a quote by your engineer and explain its meaning.
- Tell if this engineer was recognized during the time they were alive/working. If they were not, explain why.
- Based on your research, explain (in your own words) how their accomplishment(s) have benefited the world.
- Discuss anything else you learned based on your research.

7. Give a 5-min oral presentation on the famous engineer chosen in question 6. The presentation should follow the same outline as the report in question 6.

References

1. About Technical Communication, The Society for Technical Communication, https://www. tcbok.org/about-technical-communication/, viewed on June 14, 2021.
2. Jerome N. Borowick (2000), How to Write a Lab Report, Prentice Hall, ISBN: 9780130135629.
3. Diana Hopkins, Tom Reid (2020), Write Your Lab Report, SAGE Publications, ISBN: 9781529720549.
4. Kenneth G. Budiniski (2001), Engineers Guide to Technical Writing, ASM International, ISBN: 0-8780-693-8.
5. Harvey H. Hoffman (2014), The Engineering Capstone Course: Fundamentals for Students and Instructors, Springer, ISBN: 978-3-319-05896-2.
6. Silyn-Roberts, H. (2005). Professional Communications: A Handbook for Civil Engineers. United States: American Society of Civil Engineers.
7. Morley-Warner, T. 2009, Academic writing is... A guide to writing in a university context, Association for Academic Language and Learning, Sydney.
8. American Psychological Association. 2010. Publication Manual of the American Psychological Association. 6th ed. Washington, DC: American Psychological Association.
9. Bates College. 2012. "Almost everything you wanted to know about making tables and figures." How to Write a Paper in Scientific Journal Style and Format, January 11, 2012. http:// abacus.bates.edu/~ganderso/biology/resources/writing/HTWtablefigs.html.
10. Cleveland, William S. 1994. The Elements of Graphing Data, 2nd ed. Summit, NJ: Hobart Press.
11. Council of Science Editors. 2014. Scientific Style and Format: The CSE Manual for Authors, Editors, and Publishers, 8th ed. Chicago & London: University of Chicago Press.
12. University of Chicago Press. 2017. The Chicago Manual of Style, 17th ed. Chicago & London: University of Chicago Press.
13. Powerpoint tips – Simple Rules for Better PowerPoint Presentations, https://edu.gcfglobal.org/ en/, viewed June 21, 2021.
14. How to Create a Research Poster, https://guides.nyu.edu/posters, viewed June 21, 2021.
15. Gary Kroehnert (1998), Basic Presentation Skills, McGraw-Hill Australia, ISBN:9780071012102.
16. Stephan, E.A., Bowman, D.R., Park W.J., Sill B.L., Ohland, M.W. (2018), Thinking Like an Engineer: An Active Learning Approach, 4th Edition, Pearson, Upper Saddle River, New Jersey.
17. Shannon, C. E., & Weaver, W. (1949). The mathematical theory of communication. Urbana, Illinois: University of Illinois Press

18. Craig, Robert T. Communication Theory as a Field (1999). Communication Theory. 9 (2): 119–161.
19. Berlo, D. K. (1960). The process of communication; an introduction to theory and practice. New York: Holt, Rinehart and Winston. Claude E Shannon, W. W. (1949)
20. Schulz, Peter. Cobley, Paul. (2013). Theories and Models of Communication. Berlin: De Gruyter Mouton. 2013.
21. Qing-Lan Chen, Chiou-Shuei Wei, Mei-Yao Huang and Chiu-Chi Wei. (1992) A model for project communication medium evaluation and selection.
22. Chandler Daniel, The Transmission Model of Communication. http://transcriptions-2008. english.ucsb.edu/archive/courses/warner/english197/Schedule_files/Chandler/Transmission. model_files/trans.htm Archived 2019-06-24 at the Wayback Machine (1994)
23. Erik Hollnagel, David D. Woods. Joint Cognitive Systems: Foundations of Cognitive Systems Engineering. CRC Press. 2005
24. Fiske, John. Introduction to Communication Studies. London: Routledge (Chapter 1, 'Communication Theory' is a good introduction to this topic). (1982)
25. Chandler, Daniel (1994). The Transmission Model of Communication. University of Western Australia. Retrieved 2020-03-23.
26. Berlo, D. K. (1960). The process of communication. New York, New York: Holt, Rinehart, & Winston.
27. Chapter 3: How Well do I Communicate with Others?, Los Angeles Unified School District, https://achieve.lausd.net/domain/4, viewed June 23, 2021.
28. Kevin A. Smith (2004) Teamwork and Project Management. McGraw Hill Higher Education, ISBN: 9780072483123
29. Henry M. Robert (2019), Robert's Rules of Order, Blurb, Incorporated, ISBN: 9780368599347.
30. https://info.catme.org/, viewed June 23, 2021.

Part III
Engineering Fundamental Parameters

Chapter 10
Dimensions and Units

Abstract This chapter begins to introduce the concepts every engineer should know. Regardless of the type of engineering career you ultimately pursue, having a strong understanding of the fundamentals will help you solve the many different problems you will encounter throughout your academic and professional pursuits. We begin by developing an understanding of fundamental engineering dimensions and units. First, we discover why as a practicing engineer it is important that you learn to distinguish between the two terms. Then we learn to perform proper analysis with them.

By the end of this chapter, students will learn to:

- Identify physical quantities in terms of dimensions and units.
- Differentiate between fundamental and derived units.
- Recognize the different unit systems.
- Recognize the different primary, supplementary, and derived units in each system.
- What Dimensional Homogeneity is, and how to apply it to engineering problems to understand the meaning of final outcomes even before numerical solutions are calculated.
- Systematically convert units from one system to another.
- Use knowledge of dimensions and units, along with conversion rules in the solution of engineering problems.
- Correctly display units when solving engineering problems.
- Write numerical answers with correct significant digits.
- Write numerical answers with scientific notation.
- Use general techniques for estimating the unknowns of a problem.
- Recognize and report the types of reasonableness in answers to problems.

10.1   Fundamental Dimensions and Units

Throughout history, humans have learned from interacting with their environment and each other. Through observations, they realized they needed some way to describe surroundings and events to each other so that they could communicate in a

© Springer Nature Switzerland AG 2022

M. Blum, *An Inquiry-Based Introduction to Engineering*,
https://doi.org/10.1007/978-3-030-91471-4_10

consistent manner. Once common communication was established, people began to design, develop, test, and fabricate societal improvements such as tools, shelter, weapons, water transportation, and resources to grow food. In this first inquiry, you will label the categories for different physical quantities, explore the need for dimensions and units, and compare the terms.

10.1.1 **Back a very long time ago, when answering the question "how old are you?" or "how long does it take to get from here to there?" a person might answer "I am many moons old." or "it takes many moons to go from my village to your village." Based on the examples given, explain why we need physical variables.**

10.1.2 **See how many physical quantities you can name that we use in everyday life.**

10.1.3 **When baking, if we say "bake the brownies for a long time," what physical quantity are we trying to describe?**

10.1.4 **Based on the previous questions, define in your own words what the term *dimension* means.**

10.1.5 **Based on your definition to the previous question, name the seven *fundamental dimensions*. (It is from these base dimensions that all other physical quantities are defined)**

10.1.6 **When baking, we actually say "bake the brownies for *28 min*." We do not say "bake the brownies for *a long time*." Based on the two sentences, define in your own words what the term *unit* means.**

10.1.7 **Based on your definition to the previous question, list three examples of *units*.**

10.1.8 **Explain how Dimensions and Units are different from each other.**

10.1.9 **Explain how Dimensions and Units are related to each other.**

10.2   Primary and Secondary Units

10.2.1   Primary Units

Units can be likened to colors. The three primary colors (blue, yellow, red) are mixed in different combinations to make other colors (E.g., mixing blue and yellow together makes the color green). The Primary Units (also known as Base Units) are used to measure each of the fundamental dimensions. Currently, two primary unit systems are predominantly used in engineering applications throughout the world:

International System of Units (SI units): The SI unit system is the most commonly used system of measurement, with only three countries around the world not adopting its usage; the United States, Liberia, and Myanmar [1]. Essentially, it is the modern form of the metric system, built on seven base units, which correspond to the seven fundamental dimensions. The units are derived from invariant constants of nature, which are measured with extreme precision. For example, natural quantities such as the speed of light in vacuum and the triple point of water are used [2]. In addition, to specify fractions and multiples of the units, there is a set of 20 prefixes that can be applied. These prefixes are based upon multiples of ten. The SI system is the standard units used in medicine, science, engineering, and the military [3].

United States customary system (USCS or US System): The USCS is the system of measurements commonly used in the United States (US). The system evolved from the British Imperial unit system, established by the British Empire, and used during US colonial times [4]. The units derived over thousands of years from Roman, Celtic, Anglo-Saxon, and customary local units employed in the Middle Ages and were standardized by Great Britain in 1824 [5]. The US System is commonly used for measurement in commercial activities, consumer products, construction, and manufacturing because builders and manufacturers argue that measurements are easier to remember in the form of an integer number plus a fraction [6].

In order to scale numbers to primary dimensions, primary units must be assigned. The next activity provides an opportunity to determine which dimension a given unit indicates, and allows us to provide an appropriate unit given the quantity in the SI and US unit systems.

10.2.1   **Looking at Table 10.1, what is the difference between the m in row 1 and the M in row 2? (Note: one letter can represent different quantities in various engineering problems)**

10.2.2   **What does the K symbol stand for in Row 4?**

10.2.3   **Which rows have the same Base Units in both the SI system and the US system? (hint: there are three of them)**

10.2.4   **Fill in the empty cells in Table 10.1.**

Table 10.1 Table of fundamental dimensions and base units for each of the common unit systems

Row number	Primary dimension	Dimensional symbol	Base unit (SI)	Symbol (SI)	Base unit (US system)	Symbol (US system)
1	Length	L		m		
2	Mass	M		kg		lb_m
3	Time	t				
4	Temperature			K		R
5	Amount of matter (Count)	N			pound-mole	
6	Electric current			A		A
7	Amount of light (luminous intensity)	C		c		c

10.2.2   Secondary Units

Just like with colors, as the quantities we measure become more complex, so must our units. Secondary units, also known as derived units, are combinations of two or more of the primary dimensions, and thus units. Most engineering measurements are of derived dimensions in secondary units. The SI system has 22 secondary units, but there are some common ones that engineers should be familiar with. When expressing derived dimensions, it is helpful to be able to report the dimension in (1) dimensional symbols—using only the based dimensions (E.g., Length = "L") and (2) the SI and English base units. The common secondary quantities can be grouped based on Interaction Quantities, Geometric Quantities, and Rate Quantities. The next activity provides an opportunity to explore expressing derived dimensions in this way.

10.2.5 **Looking at Table 10.2, explain the reason for the three groups of quantities (Interaction versus Geometric versus Rate).**

10.2.6 **Explain the difference between Mass Flowrate and Volume Flowrate in rows 9 and 10.**

10.2.7 **In Table 10.2 row 1, what is the engineering symbol for the quantity force?**

Table 10.2 Table of Secondary Units and their derived dimensions for some common engineering quantities

Row number	Quantity	Engineering symbol	Dimensional symbols	Derived units (SI)	Derived units (USCS)
Group 1: Interaction quantities					
1	Force		$\dfrac{M \cdot L}{t^2}$		lb_f
2	Work (Energy)	E or W	$\dfrac{M \cdot L^2}{t^2}$		$ft \cdot lb_f$
3	Pressure			$\dfrac{kg}{ms^2}$	$\dfrac{lb_f}{in.^2}$
Group 2: Geometric quantities					
4	Area	A	L^2		ft^2
5	Volume				
6	Density	ρ			lb_m/ft^3
Group 3: Rate quantities					
7	Velocity	v			
8	Acceleration			$\dfrac{m}{s^2}$	
9	Mass flowrate	$\dot{m}$		$\dfrac{kg}{s}$	
10	Volume flowrate	$\dot{V}$			$\dfrac{ft^3}{s}$

10.2.8 **In Table 10.2 row 1, what is the derived SI unit of the quantity force?**

10.2.9 **In Table 10.2 row 1, what is the derived US unit of the quantity force?**

10.2.10 **In Table 10.2 row 1, explain the difference between the SI units of force and the US units of force.**

10.2.11 **In Table 10.2 row 2, based on the dimensional symbols and the US units, define the quantity *Work* in your own words.**

10.2.12 **Fill in the empty cells in Table 10.2. (Refer back to Table 10.1 for symbols and base units.)**

10.3 Dimensional Analysis and Homogeneity

To describe a physical phenomenon, engineers created mathematical formulas, which can be likened to sentences. Where the terms of the formula ($4x$, $5y^2$, etc.) equate to the words in a sentence, and the operational symbols ($+$, $\times$, $-$, $\div$) are the punctuation marks. Just like there are rules to writing sentences, there are laws that govern equations. Dimensional Homogeneity is a concept that aids engineers in observing those laws because it helps to check whether an equation of any physical phenomenon is valid and dimensionally correct or invalid. It is important to understand that all formulas in engineering must be dimensionally homogeneous and that certain *formula laws* need to be followed, namely,

1. Every term in a formula must have the same units so that arithmetic operations (such as addition, multiplication, subtraction, and division) can be carried out correctly.
2. When formula parameters are multiplied or divided, the dimensions and units are treated with the same operation rules as the numerical values they are describing.

The first inquiry allows us to discover what dimensional homogeneity means.

10.3.1 **Give three examples of substances that are homogenous.**

10.3.2 **Define what homogenous means.**

10.3.3 **Suppose we are interested in manufacturing a nail file shown in Fig. 10.1. Important quantities that define the product and how it works are listed in Table 10.3. Use the symbols in the table to develop an equation to calculate how heavy the file would be ($H =$).**

10.3.4 **Use the answer to 10.3.3 to develop an equation to calculate how heavy the file would in terms of its dimensions.**

10.3.5 **Use the answer to 10.3.4 to develop an equation to calculate how heavy the file would in terms of its units.**

10.3.6 **Is the equation that was developed in 10.3.3 through 10.3.5 valid to use to calculate how heavy the file would be? Why or why not?**

10.3.7 **Would you have to make any changes to the equation in 10.3.5 to actually calculate the heaviness of the nail file?**

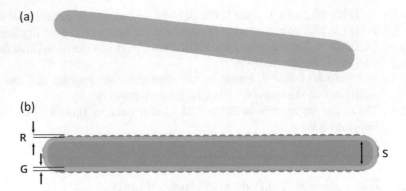

Fig. 10.1 Drawing of a nail file. (**a**) Shown together and (**b**) then broken down into some components that are labeled in Table 10.3

Table 10.3 List of quantities that affect the final product of a nail file (like the one shown in Fig. 10.1

Symbol	What it represents	Unit	Dimension
H	How heavy the file is	Newton	ML/t^2
D	Density of file grains	kg/m³	M/L^3
G	Thickness of glue to stick sand grains to the file	mm	L
R	Roughness of sand grains	mm	L
N	Amount of nail removed	in.	L
L	Length of one stroke of the nail file	cm	L
S	Thickness of the base stick	mm	L

10.3.8 **Now suppose we are interested in assessing the functionality of the nail file. Use the symbols in Table 10.3 to develop an equation to calculate how much nail particles would be removed when you ran the file over a nail once ($N =$).**

10.3.9 **Use the answer to 10.3.8 to develop an equation to calculate how much nail particles would be removed in terms of its dimensions.**

10.3.10 Use the answer to 10.3.9 to develop an equation to calculate how much nail particles would be removed in terms of its units.

10.3.11 **Is the equation that was developed in 10.3.8 through 10.3.10 valid to use to calculate how heavy the file would be? Why or why not?**

10.3.12 **Would you have to make any changes to the equation in 10.3.8 to actually calculate the heaviness of the nail file?**

10.3.13 **Using the definition you gave in 10.3.2 and the answer to 10.3.3 through 10.3.12, what does it mean to have a mathematical equation be homogeneous?**

The principle of dimensional homogeneity serves engineers because

1. It helps to check whether an equation of any physical phenomenon is valid or invalid.

2. It helps to determine the dimensions of a physical quantity or engineering constant.

This guided inquiry gives us practice using dimensional analysis.

10.3.14 Is the equation $F = mv$ valid? Explain why it is, or is not? Recall that F = force, m = mass, and v = velocity (you can check back to Table 10.1 and Table 10.2 for help).

10.3.15 Is the equation $F = ma$ valid? Explain why it is, or is not? Recall that F = force, m = mass, and a = acceleration (you can check back to Table 10.1 and Table 10.2 for help).

10.3.16 Write an expression for force in primary dimensions?

10.3.17 The surface tension of a liquid is defined as the force per unit length. Write an expression for the equation for surface tension in terms of primary dimensions only.

10.3.18 When a constant load is applied to a bar of constant cross-sectional area, as shown in Fig. 10.2, the amount by which the end of the bar will deflect can be determined from the following relationship:

$$d = \frac{PL}{AE} \quad \text{where} \quad \begin{aligned} & d = \text{end deflection of the bar} \\ & P = \text{applied load} \\ & A = \text{cross sectional area of the bar} \\ & E = \text{modulus of elasticity of the material} \end{aligned}$$

What is the primary dimension symbol(s) and SI unit for the end deflection of the bar?

10.3.19 What is the primary dimension symbol(s) and SI unit for the cross-sectional area of the bar?

10.3.20 What is the primary dimension symbol(s) and SI unit for the applied force?

10.3.21 Based on the answers to questions 10.3.18 through 10.3.20, use Dimensional Analysis to determine the modulus of elasticity in terms of primary dimensions.

Fig. 10.2 Representative example of bar being pulled by a force for use in Guided Inquiry 10.3. [Change bar to be cylinder instead of square]

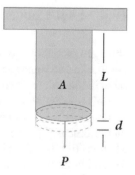

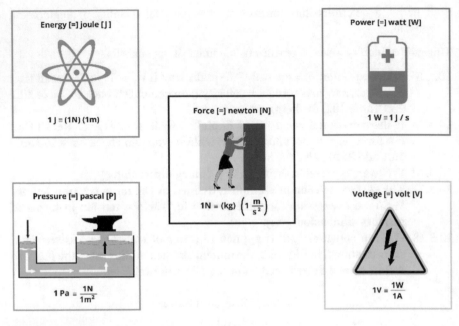

Fig. 10.3 Special names for certain derived SI units

10.3.22 **Based on the answers to questions 10.3.18 through 10.3.20, Use Dimensional Analysis to determine the units for modulus of elasticity in SI units.**

By understanding these specially named units, and using dimensional analysis, engineers can determine the dimensions of a physical quantity or engineering variable. As we measure and describe more complex phenomenon, the dimensions and units become more complex. In certain cases, some special names have been given to derived units. They are named after a famous scientist or engineer and can be found in Fig. 10.3.

10.4   Unit Conversions

Units are probably one of the most underappreciated tools in engineering existence, but they are so important. A prime example of unit importance is perfectly summarized in the article sown in Fig. 10.4.

Unit conversion is a multistep process that involves multiplication or division by a numerical factor to change the presentation of the unit, but not the value of the measurement [7]. The standard procedure for unit conversion is called the Ratio Method. In this method, we write different unit equivalences as ratios with a clear top and bottom. For example, there are 5280 ft in 1 mile, so the ratio would be set up as follows: 5280 ft/1 mile . The conversions can be between things that are also

---

NASA's metric confusion caused
Mars orbiter loss

(CNN) -- NASA lost a $125 million Mars orbiter because one engineering team used metric units while another used English units for a key spacecraft operation, according to a review finding released Thursday.

For that reason, information failed to transfer between the Mars Climate Orbiter spacecraft team at Lockheed Martin in Colorado and the mission navigation team in California. Lockheed Martin built the spacecraft.

"People sometimes make errors," said Edward Weiler, NASA's Associate Administrator for Space Science in a written statement.

"The problem here was not the error, it was the failure of NASA's systems engineering, and the checks and balances in our processes to detect the error. That's why we lost the spacecraft."

Fig. 10.4 Text from print article about Mars Climate Orbiter from CNN September 30, 1999 [10]

exchangeable. For example, if a car gets 25 mpg, that means the car can drive 25 miles using 1 gallon of gas. The equivalency is 25 miles/1 gal .

To do the conversion, choose the ratio that will allow the undesired unit or units to cancel out. Units will cancel when they appear once on top and once on bottom in a multiplication string. Note that it may be necessary to multiply by more than one conversion ratio in more difficult problems. In that case, you set up multiple ratios of the units and cancel them out until only the desired unit remains. Also, it helps to remember that any number can be thought of as the ratio of that number over one. As practicing engineers, when performing analysis, you will typically use unit conversion for two purposes: (1) converting within a given unit system and (2) from one system of units to another. This guided inquiry gives us practice doing both.

10.4.1 **Write a ratio to show how many kilometers are there in a meter?**

10.4.2 **Using the ratio method, calculate how many kilometers are in 32.4 m?**

10.4.3 **If your car gets the mileage described in the above paragraph, how far can you drive on a full tank of gas, if your car has an 18 gal tank?**

10.4.4 **If you can run a mile in 8.5 min, how long does it take you to run 26,400 ft?**

10.4.5 **Using the answer from 10.4.4, how many miles have you run?**

10.4.6 **A honeybee's wing flap is said to take about 5 milliseconds. How many times does a honeybee flap its wings in 1 week?**

10.5 Significant Digits

After going through the process of unit conversion or solving a set of calculations, or measuring a physical quantity, engineers need to record the numerical results of these tasks. They record the results of measurements and calculations using numbers. Rarely do engineers deal with "basic" numbers, because the numbers we use are typically due to measurements, and there is no perfectly exact measurement. Therefore, all measurements, and thus engineering calculations, have uncertainty. So as engineers, we have to decide where to stop displaying the values. For example, the value Pi (π) is what mathematicians call an "infinite decimal," meaning the digits after the decimal point go on forever. Pi has been calculated to be as high as 31.4 trillion digits [8]. It can be displayed as 3.14, 3.1415926... and 3.1415926535 8979323846.... So based on the example of Pi, where is the proper spot to stop a value?

Where the number display gets stopped or truncated is called the significant digit. *Significant digits* (also known as *significant figures* (*SF*)) express the extent to which computed data or a recorded number is dependable. Engineers use significant figures to communicate the level of uncertainty in measurement or calculation. Listed below are several rounding procedures that help scientists, engineers, and manufacturers regularize technical documents, specifications, and other unit applications.

- For data measurements, we record the accuracy of measurements to the "least count." The *least count* is one-half of the smallest scale division of the measuring instrument.
- Nonzero digits are always significant.
- Any zeros between two significant digits are significant.
- "Point right, not left"—This means that if the number has a decimal point, first move to the right. If the number does not have a decimal point, move to the left.
- Moving in the indicated direction, start counting with the first *nonzero* digit, and keep counting until there are no more digits in that direction.
- Exact values have infinite SF. Values are exact if they are simple counts (the dining room table has seven chairs) or by definition (1 m = 1000 mm exactly because we defined it that way).

This inquiry lets us practice counting and displaying significant digits correctly.

10.5.1 **In Fig. 10.5a, what units is the ruler measuring?**
10.5.2 **In Fig. 10.5a, how many divisions make up 1 inch?**
10.5.3 **Based on question 10.5.2, what is each division on the rule in Fig. 10.5a?**
10.5.4 **Based on question 10.5.3, what is the accuracy of the ruler?**
10.5.5 **When sighting a measurement, it is typically recorded as the value ± the least count. In Fig. 10.5a, how would you record the measurement where the red arrow is pointing?**

(a) Ruler

(b) thermometer measuring

(c) mass scale measuring

(d) Electricity meter measuring

Fig. 10.5 Images that will be used for Guided Inquiry 10.5. Measurements taken with (**a**) ruler, (**b**) thermometer, (**c**) mass scale, and (**d**) electricity meter

10.5.6 For each of the values in Fig. 10.5, list the number of significant figures, least count, the decimal place of the uncertainty, and the number in that decimal place.

Significant figures will also appear in calculations. Moreover, since engineers use measurements a lot for calculations, that means the values we get from calculations also have uncertainty. There are also rules for significant digits when performing calculations.

For *addition and subtraction*, use the following rules:

- Look at the number of significant figures in the decimal portion ONLY of each number in the problem. The number with the least decimal places is called the **Limiting Term**.
- Add or subtract in normal style.
- Your final answer must have the same number of decimal significant figures as the limiting term.

For *multiplication and division*, use the following rules:

- The fewest decimal places in any number of the calculation determines the number of significant figures in the answer. You are now looking at **the entire number**, not just the decimal portion.

If you are using a combination of mathematical operations (+, ×, −, ÷), you apply the significant figure rules according to order of mathematical operations. AND always remember, regardless of whether you are adding, subtracting, multiplying, or dividing, **WAIT TO ROUND THE FINAL ANSWER UNTIL THE END OF THE PROBLEM!!**

This inquiry lets us practice performing calculations and displaying appropriate significant digits.

10.5.7 **How many SF does 162.47 have?**

10.5.8 **How many SF does 5 have?**

10.5.9 **How many SF does 6.2 have?**

10.5.10 **If your calculator added the numbers 162.47 + 5 together, what would be displayed as the answer?**

10.5.11 **Using the addition and subtraction rules of SF, which number in question 10.5.10 has the fewest decimal places? How many SF does the number have?**

10.5.12 **Using the answers to questions 10.5.10 and 10.5.11, how should the answer to the question 162.47 + 5 be displayed in order for it to have the correct number of significant figures?**

10.5.13 **Using the addition and subtraction rules of SF, what is the result of 147.333 − 6.2?**

10.5.14 **If your calculator divides the numbers 162.47 ÷ 6.2 together, what would be displayed as the answer?**

10.5.15 **Using the multiplication and division rules for SF, which number in question 10.5.14 has the least number of significant figures? How many SF does the number have?**

10.5.16 **Using the answers to questions 10.5.14 and 10.5.15, how should the answer to the question 162.47 ÷ 6.2 be displayed in order for it to have the correct number of significant figures?**

10.5.17 **If you were adding 7.0 inches to 3.5 ft, how you would perform the conversion? (You just need to explain a procedure, you do not need to use numbers yet, and there is not just one way to do it).**

10.5.18 **Add 7.0 in. to 3.5 ft and display with correct SF.**

Lastly, when displaying the final measurement or calculation, engineers understand that the solution will not be exact. Therefore, we round the final significant digit to an easily displayable value that is less exact but has about the same value as the more detailed number. A number can be rounded to any place value that the engineer chooses. Here are the general rules for rounding

- If the number you are rounding is followed by a value *less than 5* (namely, 0, 1, 2, 3, or 4), round the previous number *down*.
- If the number you are rounding is followed by a number *equal to or greater than 5* (namely 5, 6, 7, 8, or 9), then round the previous number *up*.

This inquiry lets us practice rounding.

10.5.19 **What was the answer to questions 10.5.10?**

10.5.20 **In question 10.5.19, label all the numbers and their decimal positions.**

10.5.21 **If you needed to round the answer of 10.5.20 to the nearest tenth, what decimal place would you be checking before you round?**

10.5.22 **What is the number in that decimal place?**

10.5.23 **Based on your answer to question 10.5.22, would you round up or down?**

10.5.24 **What would the number in question 10.5.19 be rounded to if you were rounding to the nearest tenths place?**

10.5.25 **What would the number in question 10.5.19 be rounded to if you were rounding to the nearest ones place? Tens place? Hundreds place?**

10.6 Scientific Notation

Scientific notation is a special way to write numbers. Typically, it is used with very small or very large numbers. Numbers are written as multiples of a power ten. This base ten number system is commonly used in math, science, and engineering because it has several advantages

- It makes it easy to display the appropriate number of significant figures.
- It makes writing very large or very small numbers easier and clearer.
- It allows for less error performing calculations with very large or very small numbers.

To display any number using scientific notation, write the number in the form of raising it to the power of ten, $N \times 10^m$, where N is a set of digits (integers that may or may not include a decimal point), and "m" is an integer that is any whole number, positive or negative. To write the number in its regular form, the integer "m" indicates how many times the decimal point should be moved to the right (if m is positive) or to the left (if m is negative). When a number is properly written in scientific notation, the value of m is the "order of magnitude" of that number. In addition, all of the digits in N are considered to be "significant," regardless of the order of magnitude. Calculators or computer programs sometimes replace the $\times 10$ by the term "E". (E.g., $3.12E4$ is the same as 3.12×10^4). This inquiry lets us practice displaying numbers using scientific notation.

10.6.1 **How many times do you need to multiple 10 to itself to get 1000?**

10.6.2 **What is the index of power (how many times you use the number ten in multiplication) for 1000?**

10.6.3 **Write the number 1000 in powers of ten form.**

10.6.4 **Write the number seven thousand in number form.**

10.6.5 **Using the answers to questions 10.6.1 through 10.6.4, write the number seven thousand in scientific notation.**

10.6.6 **Compare the answers to questions 10.6.4 and 10.6.5 (similarities and differences).**

10.6.7 **Write the number 4,900,000,000 in scientific notation.**

10.6.8 **Write the number 0.00000451 in scientific notation.**

10.6.9 **What is the order of magnitude for questions 10.6.7 and 10.6.8?**

The SI units system has a series of prefixes and symbols of decimal multiples to indicate the order of magnitude of a base unit. A list of prefixes, symbols, multiples, and order of magnitudes is shown in Table 10.4. It is good for engineers to memorize the highlighted prefixes in the table.

10.7   Estimation and Reasonableness

Estimation is an important skill that engineers need to develop. It is the process of finding a rough calculation of the value, number, quantity, or extent of something that is uncertain. Often in our profession, we need to solve problems for which we do not have all the information. It is also the final step in the problem-solving process, so engineers need to be adept at identifying and making estimates for the critical missing information and be able to recognize if their solution is in the "right ballpark" or *reasonable*. Estimation is a vital part of setting up a problem, and there is no quick and simple answer to become skilled at estimation. It takes practice and experience. However, there are a few standard estimation tricks that engineers use depending on the type of problem that needs to be solved [9].

Trick 1: **Simplify Geometry**: Many problems will have complex parts and forms. Often times these complex profiles can be simplified into simple shapes, areas, and volumes, which we know the equations for.

Table 10.4 List of symbols, prefixes, multiplication factors, and orders of magnitude for SI units

Prefix	SI symbol	Multiplication factor	Order of magnitude
yotta	Y	10^{24}	24
zetta	Z	10^{21}	21
exa	E	10^{18}	18
peta	P	10^{15}	15
tera	T	10^{12}	12
giga	G	10^{9}	9
mega	M	10^{6}	6
kilo	K	10^{3}	3
hecto	H	10^{2}	2
deka	Da	10^{1}	1
deci	D	10^{-1}	−1
centi	C	10^{-2}	−2
milli	M	10^{-3}	−3
micro	M	10^{-6}	−6
nano	N	10^{-9}	−9
pico	P	10^{-12}	−12
femto	F	10^{-15}	−15
atto	A	10^{-18}	−18
zepto	Z	10^{-21}	−21
yocto	Y	10^{-24}	−24

Trick 2: **Use Analogies**: For this strategy, you can estimate something unknown by comparing it to known quantities (sizes, weights, theories, etc.)

Trick 3: **Aggregate**: You can estimate the quantity of something by estimating one, then scaling up and adding up the parts.

Trick 4: **Bound the Solution**: Place an *upper* limit and a *lower* limit on the answer. It will help you to estimate your solutions within reasonable physical bounds.

Trick 5: **Extrapolate from Samples/Models**: For very complex problems, models, statistics, or sample data may be used to estimate a solution.

The next inquiry gives us practice utilizing the estimation tricks.

10.7.1 **What simple shape does the human head most closely resemble?**

10.7.2 **What simple shape does the human torso most closely resemble?**

10.7.3 **What simple shape do the individual human arms and legs most closely resemble?**

10.7.4 **Using the answers to 10.7.1 through 10.7.3, explain a procedure of how you would estimate the surface area of a person.**

10.7.5 **Using 10.7.4, write an equation for the estimation of the surface area of a person.**

10.7.6 **What is the definition of density?**

10.7.7 **What is the density of water?**

10.7.8 **What happens to people when they go swimming? Do they sink or float?**

10.7.9 **Based on question 10.7.8, since people are mostly water, what analogy can we make about the density of people?**

10.7.10 **Using questions 10.7.6 through 10.7.9, explain a procedure to estimate the volume of a person.**

10.7.11 **Nominally, how much does a slice of pizza cost?**

10.7.12 **Nominally, how many times a week does a typical undergraduate college student buy pizza?**

10.7.13 **Use the answers to questions 10.7.11 and 10.7.12 to explain a procedure to estimate the amount of money all students at your school spend on pizza each year at college.**

Once the solution of a problem is estimated, it is also the responsibility of the engineer to analyze the *reasonableness* of the answer. This means that we need to explain the quality and appropriateness of the solution. There are two main types of reasonableness in answers that engineers are concerned about two things: (1) physical reasonableness and (2) reasonable precision.

1. Is the answer **physically reasonable?**

Here, we ask ourselves does the answer makes physical sense based on our understanding of the situation. To check whether an answer is physically reasonable see the following checklist:

- If the final answer is in units that you do not intuitively understand, convert to units you are familiar with.
- If the solution is a mathematical model, check the behavior of the model at extremal values (namely, very small and large numbers).
- Finally, check to see if the answer makes sense in the physical world.

Let us look at a guided inquiry that investigates the concept of physical reasonableness.

10.7.14 **If someone said that they throw a baseball 50 million millimeters per minute, could you tell if this was a reasonable speed? Explain Why or Why not?**

10.7.15 **Explain a procedure you would use to check if the claim in question 10.7.14 was reasonable.**

10.7.16 **Is the claim that someone can throw a baseball 50 million millimeters per minute physically reasonable? Explain your answer.**

2. Is the answer **reasonably precise?**

Based on the parameters in the problem, we need to ask ourselves is the number answer *accurate* and *precise*? For an engineer to understand whether an answer is reasonably precise, it is important that they understand the difference between the terms accurate, repeatable, and precise.

Accurate is used to describe how close a measurement (or calculation) is to a "true" parameter. In engineering, some standard parameters are compared for accuracy. (1) A theoretical calculation can be compared with experimental measurement. (2) Two measurements can be compared to each other if they were taken with two different measurement techniques or devices. (3) Two theoretical calculations of the same factor, using different theories, can be compared. The difference between the true value and the other value is called the error. The percent (%) error can be measured using the following equation:

$$\%\text{Error} = \left| \frac{\text{True value} - \text{Other value}}{\text{True value}} \right| \times 100\%$$

For measurements, you can also take the average of the sampled data and compare it to the true value to check for accuracy and % error.

Repeatable is used to describe how close multiple measurements of the same value are to each other. It does not consider whether they are close to the "true" value or not.

Precision is a combination of accurate and repeatable. The number of significant figures reported in the answer reflects precision. The more significant figure reported, the more precise the value is.

The next guided inquiry lets us explore in more depth the meaning of these terms.

10.7.17 Looking at Fig. 10.6, do any of the data points in Sample A land on the true temperature (bulls eye)?

Fig. 10.6 Visual representation of probes measuring the temperature of a hot tub (True Temperature = 102 °F, probes are pink circles)

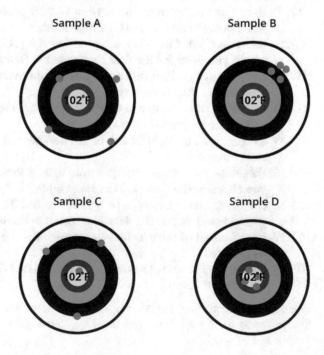

10.7.18 **Based on your answer to question 10.7.17, looking at Fig. 10.6, would you characterize Sample A as accurate?**

10.7.19 **Looking at Fig. 10.6, do any of the data points in Sample A land near each other?**

10.7.20 **Based on your answer to question 10.7.19, looking at Fig. 10.6, would you characterize Sample A as repeatable?**

10.7.21 **Based on your answers to questions 10.7.17 through 10.7.20, how would you describe the data illustrated in Sample A?**

10.7.22 **Looking at Fig. 10.6, do any of the data points in Sample B land on the true temperature (bulls eye)?**

10.7.23 **Based on your answer to question 10.7.22, looking at Fig. 10.6, would you characterize Sample B as accurate?**

10.7.24 **Looking at Fig. 10.6, do any of the data points in Sample B land near each other?**

10.7.25 **Based on your answer to question 10.7.24, looking at Fig. 10.6, would you characterize Sample B as repeatable?**

10.7.26 **Based on your answers to questions 10.7.22 through 10.7.25, how would you describe the data illustrated in Sample B?**

10.7.27 **Looking at Fig. 10.6, do any of the data points in Sample C land on the true temperature (bulls eye)?**

10.7.28 **Based on your answer to question 10.7.27, looking at Fig. 10.6, would you characterize Sample C as accurate?**

10.7.29 **Looking at Fig. 10.6, do any of the data points in Sample C land near each other?**

10.7.30 **Based on your answer to question 10.7.29, looking at Fig. 10.6, would you characterize Sample C as repeatable?**

10.7.31 **Based on your answers to questions 10.7.27 through 10.7.30, how would you describe the data illustrated in Sample C?**

10.7.32 **Looking at Fig. 10.6, do any of the data points in Sample D land on the true temperature (bulls eye)?**

10.7.33 **Based on your answer to question 10.7.32, looking at Fig. 10.6, would you characterize Sample D as accurate?**

10.7.34 **Looking at Fig. 10.6, do any of the data points in Sample D land near each other?**

10.7.35 **Based on your answer to question 10.7.34 looking at Fig. 10.6, would you characterize Sample D as repeatable?**

10.7.36 **Based on your answers to questions 10.7.32 through 10.7.35, how would you describe the data illustrated in Sample D?**

10.7.37 **Based on all of the previous questions, which Sample would you label as precise?**

10.7.38 **Using the answers to questions 10.7.17–10.7.37, fill in the last three columns of Table 10.5.**

Finally, when reporting engineering calculations or measurements, you should always report them such that there is an implied higher level of accuracy than is

Table 10.5 Displaying the values measured by the probes in the hot tub (True Temperature = 102 °F)

	Probe #1 (°F)	Probe #2 (°F)	Probe #3 (°F)	Probe #4 (°F)	Accurate (Y/N)	Repeatable (Y/N)	% Error
Sample A	90	80	83	87			
Sample B	91	90	89	90			
Sample C	102	95	79	109			
Sample D	102	100	102	101			

actually known. This means that you use the fewest number of decimal places without reducing the usefulness of the answer. This final guided inquiry lets us practice this concept.

10.7.39 **What is the equation for the area of a circle?**

10.7.40 **The value Pi (π) is an "infinite decimal." Describe in your own words what that means.**

10.7.41 **If we measure the radius of a circle to be 2.63 cm. How many decimal places of accuracy did we measure out to?**

10.7.42 **If you calculate the area of a circle with a radius that is in 10.7.41 with your calculator, what does it report?**

10.7.43 **Based on the answers to questions 10.7.39 through 10.7.43, what is the proper way to display the answer?**

10.7.44 **Now we measure another circle with a new instrument and we measure the radius as 0.000024 cm. If you calculate the area of a circle with this new radius, what does your calculator report?**

10.7.45 **For question 10.7.44 what is the proper way to display the answer?**

End of Chapter Problems

IBL Questions

IBL1: An important tool that engineers need to develop is the ability to estimate physical values; sizes, weights, distances, etc. Sometimes estimation needs to occur without an actual measurement tool. To develop a sense of physical quantities, use the table given and estimate the values for the objects provided. Then measure or look up the actual dimensions. Finally, calculate the percent error of your estimations.

Physical objects	Estimated values		Actual measurements		% Error	
	(in.)	(m)	(in.)	(m)	(in.)	(m)
Cell phone area						
Door height						
Basketball hoop height						
Thickness of credit card						
Length of football field	(yards)	(meters)	(yards)	(meters)	(yards)	(meters)
Amount of liquid in a milk carton	(gal)	(L)	(gal)	(L)	(gal)	(L)
Temperature of your dorm room	(°C)	(°F)	(°C)	(°F)	(°C)	(°F)

IBL2: Use dimensional analysis and homogeneity to answer the following questions:

1. Which of the following equations is dimensionally homogeneous: $F = mV^2/R$ or $F = mV^2/t$?
   (Note: F = force, m = mass, V = velocity, R = radius, and t = time)
   (can be solved using SI units or dimensions)
2. Are the following equation's definition correct? Explain why. (Note: m = mass, A = area, V = velocity, a = acceleration, ρ = density, and t = time)

   - $V = m^{1/3}/t \cdot \rho^{1/3}$
   - $t = A^{1/2} \cdot V$
   - $a = V^2/A^{1/2}$

IBL3: Use scientific notation and estimation techniques to answer the following questions:

1. Express the following values in scientific notation

   - 0.000000000000003 s
   - 2,000,000,000,000,000,000,000 bytes

2. Given the units for the values in the previous question, use the appropriate prefix and write the unit label.
3. Explain how you would estimate the mass of an elephant.

4. Estimate the mass of an elephant.
5. Compare your solution to data that you look up on elephant's mass.
6. Discuss how close was your estimate to the published values.
7. Estimate how many basketballs it would take to completely fill your universities stadium from bottom to top. Explain your process and compare your value with your partner, team, or the class.

Practice Problems

1. Looking at Fig. 10.2 we see that force can also be displayed in Newton [N]. Redo question 10.3.22 to see what the units of E would be if the load was expressed in Newtons.
2. Looking at Table 10.2 Row 3, based on the USCS units, what would be the definition of the quantity Pressure?
3. For the given SI to US conversion statements, prove that they are correct or incorrect.

   - $35 \text{ m/s}^2 = 120 \text{ ft/s}^2$
   - $500 \, N = 112 \text{ lb}_f$
   - $200 \text{ kPa} = 29 \text{ lb}_f/\text{in}^2$
   - $150 \text{ kg} = 300 \text{ lb}_m$
   - $2000 \text{ kg/m}^3 = 125 \text{ lb}_m/\text{ft}^3$
   - $50 \text{ m}^3 = 2000 \text{ ft}^3$

4. For the given US to SI conversion statements, prove that they are correct or incorrect.

   - $50 \text{ lb/in}^2 = 500 \text{ kPa}$
   - $75 \text{ miles/h} = 200 \text{ km/h}$
   - $300 \text{ lb}_f = 500 \text{ N}$
   - $250 \text{ lb}_m = 113 \text{ kg}$

5. For the following numbers, state the number of significant digits

   - 846.8
   - 3.4×10^3
   - 0.00456

6. For the following mathematical statements, fully solve and display the correct number of significant digits

   - $2.38495 + 11.2$
   - 2447×13
   - $4.646 \div 14$
   - $160 - 0.631$
   - $160 - 0.6$

References

1. The World Factbook Appendix G. CIA. Retrieved February 18, 2019 https://www.cia.gov/library/publications/the-world-factbook/appendix/appendix-g.html
2. Materese, Robin (November 16, 2018). "Historic Vote Ties Kilogram and Other Units to Natural Constants". NIST. Retrieved February 18, 2019 https://www.nist.gov/news-events/news/2018/11/historic-vote-ties-kilogram-and-other-units-natural-constants
3. The World Factbook Appendix G. CIA. Retrieved February 18, 2019 https://www.cia.gov/library/publications/the-world-factbook/appendix/appendix-g.html
4. Comings, E. W. (1940). "English Engineering Units and Their Dimensions". Ind. Eng. Chem. 32 (7): 984–987. doi:https://doi.org/10.1021/ie50367a028.
5. British Imperial System, Encyclopaedia Britannica, retrieved February 18, 2019 https://www.britannica.com/science/British-Imperial-System.
6. Ed Tenner, (May 2005). "The Trouble with the Meter" Technologyreview.com reviewed February 18, 2019 https://www.technologyreview.com/s/404016/the-trouble-with-the-meter/.
7. "unit conversion" https://www.nist.gov/pml/weights-and-measures/metric-si/unit-conversion. Viewed March 8, 2019.
8. "Even After 31 Trillion Digits, We're Still No Closer To The End Of Pi", https://fivethirtyeight.com/features/even-after-31-trillion-digits-were-still-no-closer-to-the-end-of-pi/, viewed March 14, 2019.
9. Mark Thomas Holtzapple, W. Dan Reece (2000), Foundations of Engineering, McGraw-Hill, New York, New York, ISBN: 978-0-07-029706-7.
10. NASA's metric confusion caused Mars orbiter loss, September 30, 1999, Web posted at: 1:46 p.m. EDT (1746 GMT), http://www.cnn.com/TECH/space/9909/30/mars.metric/index.html

Chapter 11
Length

Abstract Over the next 4 years of studying, an important part of learning the fundamentals of engineering is developing an awareness of surroundings, and how they influence the design or solution of a problem. In this chapter, you will explore the role of the fundamental dimension of length. You will see its importance in engineering application, and learn about other length-related variables. The concepts and ideas introduced in this chapter will reappear throughout your engineering curriculum, as you study more specific concepts in detail.

By the end of this chapter, students will learn to:

- Explain the importance of the notion of length in engineering analysis.
- List examples of length and related variables.
- Develop relationships between units of length in multiple unit systems.
- Define a dimensionless parameter, explain the significance of dimensionless parameter, and describe some length dimensionless parameters.
- Explain and review trigonometric concepts and functions.
- Convert between the rectangular coordinate system and two other coordinate systems.
- Describe the role of area in engineering, as well as its units.
- Using multiple methods, calculate and measure simple and composite areas.
- Describe the role of volume in engineering, as well as its units, calculate and measure volumes.
- Explain what is meant by second moment of area and describe its role, and calculate simple shapes moment of area.

11.1 Importance of Length Parameters

As we learned in Chap. 10, humans established dimensions to be able to describe the physical world around them. One of the most widely used dimensions in engineering is length. It is used in every discipline of engineering and for many applications. This guided inquiry lets us explore the significance of length and identify its importance in many aspects of our lives and engineering decisions.

11.1.1 **When constructing a classroom (like the one you are in), describe the important design factors that have to do with length.**

11.1.2 **When designing a cell phone, describe important design factors that have to do with length, and explain why they are important.**

11.1.3 **In your own words, what is the definition of *length*?**

11.1.4 **Explain why the concept of length is important in engineering.**

11.1.5 **Give examples of length and length-related variables in everyday life.**

11.1.6 **Give me some descriptions of length.**

11.1.7 **Explain the weakness with the descriptions given in the previous question.**

11.2   Length Units

The last question in the previous guided inquiry showed us that there are many ways to describe the dimension of length but to be exactly prescriptive we need to have length divisions, which we know as units. The next guided inquiry helps us to investigate the length units of the US Customary system.

11.2.1 **List the most common length units for the US system, starting with the smallest division and going to the largest.**

11.2.2 **Match each of the answers listed above to estimate the size of the following objects: width of a quarter, length of a man's shoe, length of a guitar, and length of an airportrunway.**

11.2.3 **To convert larger units to smaller units, multiply the number of larger units by the conversion factor for the appropriate smaller units. How many inches are in 1 ft? How many feet are in 1 yard?**

11.2.4 **To convert smaller units to larger units, divide the number of smaller units by the conversion factor for the appropriate larger units. How many feet are in 1 in.? How many yards are in 1 ft?**

11.2.5 **Using the strategies and answers to the previous two questions, fill out Table 11.1 for US unit length conversions.**

The SI base unit for length is the *meter* (*m*). Then multiples and fractions of the meter is used in order the keep the numbers manageable and give the dimensions a relatable scale. The next guided inquiry helps us to investigate the length units of the SI system and explore how to convert within the system.

11.2.6 **List the most common length units for the SI system, starting with the smallest division and going to the largest.**

11.2.7 **Match each of the answers listed above to estimate the size of the following objects: size of a sugar molecule, paper clip thickness, fingernail width, length of a guitar, length of 10 football fields.**

11.2.8 **To convert larger units to smaller units, multiply the number of larger units by the conversion factor for the appropriate smaller units. How**

Table 11.1 Summary of US system units of length

Inches	Feet	Yards	Miles
1			$\frac{1}{63,360}$
	1		
		1	$\frac{1}{1760}$
63,360		1760	1

Table 11.2 Summary of SI units of length

Millimeter	Centimeter	Meter	Kilometer
			1
		1	
10	1		
1			$\frac{1}{1,000,000}$

many centimeters are in 1 m? How many meters are in 1 km? (hint: look back to Table 10.3 for SI unit prefixes and orders of magnitude)

11.2.9 **To convert smaller units to larger units, divide the number of smaller units by the conversion factor for the appropriate larger units. How many meters are in 1 mm? How many kilometers are in 1 m?**

11.2.10 **Using the strategies and answers to the previous two questions, fill out Table 11.2 for SI unit length conversions.**

Depending on where and how you were raised, you will develop a feel for either the US or SI units system. You will intuitively recognize how large something is in one system. To convert between the two systems, it is not terribly important to know all the conversions. It is important to recall the unit conversion process and then memorize a few key conversions. It is best when the numbers are familiar items. The next guided inquiry gives us practice recognizing familiar length items and converting between unit systems.

11.2.11 **What is the name of the length measurement instruments in Fig. 11.1? How can you tell the units?**

11.2.12 **Looking at Fig. 11.1b, what is the unit system of measurement?**

11.2.13 **Looking at Fig. 11.1b, what is the fundamental dimension for that measurement?**

11.2.14 **Looking at Fig. 11.1b, develop a conversion statement in the given unit system.**

11.2.15 **Looking at Fig. 11.1a, what is the unit system of measurement?**

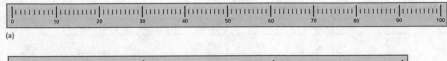

(a)

(b)

Fig. 11.1 Length measurement instruments

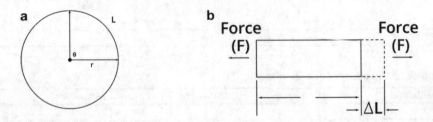

Fig. 11.2 (**a**) The relationship between the angle (θ) in radians, the length of the arc (L), and the radius of the arc (r). (**b**) A bar subjected to a pulling load leads to deformation (ΔL) from original length (L) which induces strain (ε)

11.2.16 **Looking at Fig. 11.1a, what is the fundamental dimension for that measurement.**

11.2.17 **Looking at Fig. 11.1a, develop a conversion statement in the given unit system.**

11.2.18 **Compare Fig. 11.1a and b, develop a conversion between SI and US units (it does not have to be precise at this time).**

Finally, there are some length variables that have "dimensionless units." Even though the term seems counterintuitive (how can you have dimensions that do not have units associated with them?) it is possible. Certain length terms can be defined as a ratio of physical constants, or properties, such that the dimensions cancel out. Two important unit-less length terms that you will continually encounter throughout your engineering study are Radians and Strain. *Radians* are the SI unit of an angle, defined as the ratio of the length of the arc length circle created by the angle, divided by the radius of the circle. *Strain* is when a system responds to applied stress by deforming. It is defined as the amount of deformation in the direction of applied stress divided by the initial length of the material. This next inquiry helps us to better understand these two unit-less length parameters.

11.2.19 **What is the formula for the circumference of a circle?**

11.2.20 **Looking at Fig. 11.2a, how much of the circle does the arc L transverse?**

11.2.21 **Looking at Fig. 11.2a, write an equation for arc length L in terms of the circumference of the circle?**

11.2.22 **Based on the definition of a radian, write the formula to measure the angle θ.**

11.2.23 **Using the answers to questions 11.2.19 through 11.2.22 calculate the angle θ in Fig. 11.2a in radians. What are the units for this calculation?**

11.2.24 **If the radius of the circle is 7 cm, what is the angle θ?**

11.2.25 **If the radius of the circle is 10 cm, what is the angle θ?**

11.2.26 **Are the answers to questions 11.2.24 and 11.3.25 different or the same? Explain why.**

11.2.27 **Looking at Fig. 11.2b and based on the definition of radian you derived above, develop an equation for strain (ε).**

11.2.28 **Looking at Fig. 11.2b, if the original length of the bar was 10 in. and the force stretched the bar to go another 5 in., calculate the strain seen in the bar. What are the units for the calculation?**

11.2.29 **Based on your answer to questions 11.2.19 through 11.2.28, why are both radians and strain called "unit-less"? Are they really "unit-less"?**

11.3 Length Calculations

11.3.1 Trigonometry Review

One prime example of important length calculations that all engineers need to understand and utilize is trigonometry. Trigonometry is the mathematical study of the relationships between sides and angles of triangles. These relationships are called trigonometric functions and are pervasive in the many math, science, and engineering studies including applied geometry, geodesy, surveying, dynamics, celestial mechanics, navigation, and solid mechanics [1]. The next guided inquiry helps us review trigonometric concepts and functions that will be important throughout your engineering career.

11.3.1 **What is the shape in Fig. 11.3a?**

11.3.2 **Explain the Pythagorean relationship between the sides of a right triangle.**

11.3.3 **What are the three main functions in trigonometry that are used with a right triangle?**

11.3.4 **Looking at Fig. 11.3a and using the answers to question 11.3.3, define the three main functions in right triangle trigonometry.**

11.3.5 **Using the answer to question 11.3.4, develop a plan for how to calculate how tall a tree was such as in Fig. 11.3b without climbing the tree. You have a protractor to measure angles and you tape a drinking straw to the protractor to look through.**

11.3.6 **Using the answer to question 11.3.5, If you measured an angle of 35° and were standing a distance of 20 ft away from the tree, how tall would you calculate the tree to be?**

11.3.7 **Looking at the triangles in Fig. 11.3c, do they have right angles?**

11.3.8 **What are the shapes in Fig. 11.3c?**

11.3.9 **Looking in Fig. 11.3c triangle #1, is the angle c small or large?**

11.3.10 **Looking in Fig. 11.3c triangle #1, is line C small or large?**

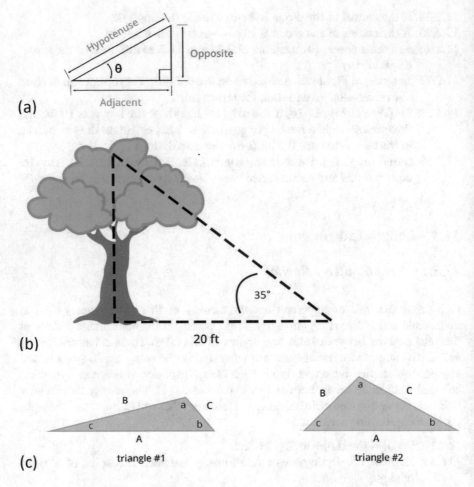

Fig. 11.3 (a) Right Triangle with defined lengths and angle θ. (b) Example of using trigonometry to discover height. (c)Two scalene triangles with all different sides and angles

11.3.11 **Looking in Fig. 11.3c triangle #2, is the angle c small or large?**

11.3.12 **Looking in Fig. 11.3c triangle #2, is line C small or large?**

11.3.13 **Using questions 11.3.9 through 11.3.13, explain the relationship between the angle C and side c for both triangles?**

11.3.14 **For triangles #1 and #2, does this relationship hold true for the other sets of angles and sides (namely a, A and b, B)?**

11.3.15 **If we say that the relationship for each set is the ratio of the length of the side over the sine of the angle, write out full relationship for each set (also known as the law of sines).**

11.3.16 **Using the answer to question 11.3.15, if you had triangle #1, how many knowns would you need to solve for the length of side b? Give an example of the knowns.**

11.3.17 If triangle #1 had length $b = 5.7$ cm, angle $B = 79°$, and angle $C = 32°$, how long is side c?

11.3.18 Looking at triangle #2, if you are given a value for side a, side c, and the angle B, can you solve for the length of side b using the law of sines you just defined in question 11.3.15?

11.3.19 Looking at triangle #2, if you are given a value for side a, side b, and side c, can you solve for angle B using the law of sines you just defined in question 11.3.15?

11.3.20 Explain another law from trigonometry that will solve the problems posed in questions 11.3.18 and 11.3.19.

11.3.21 If triangle #2 has side $a = 9$ in., side $b = 5$ in., and side $c = 8$ in., what is angle C?

It is very likely that you have used trigonometric tools to analyze some problems in the past. It is also possible that you have not used them recently. Therefore, these tools need to be taken out of your mental tool kit and dusted off, particularly Pythagorean relation, SOH-CAH-TOA, sine, and cosine rules. Some other useful trigonometric identities can be found in Fig. 11.3d that you should be familiar with and be able to apply to future engineering problems.

11.3.2 Coordinate Systems

Another example of important length calculations is coordinate systems. They are used for locating things with respect to an origin. You use coordinate systems every day in life. How long does it take for me to roll out of bed from my dorm room and get to class? Your dorm room is the origin. A coordinate system is a scheme that allows us to identify any point in a plane or in three-dimensional space by a set of numbers. There are many different types of coordinate systems. Based on the nature of a problem, an engineer may use one coordinate system over another. Students are typically most familiar with using rectangular (aka Cartesian) coordinates. In rectangular coordinates, distances are interpreted as numbers, essentially as the lengths of the sides of a rectangular "box." For example, to locate an object at point A in Fig. 11.4, with respect to the origin (point 0) of the Cartesian system, you move along the x-axis by x amount (or unit steps) and then you move along the dashed line parallel to the y-axis by y amount. Finally, you move along the dashed line parallel to z-direction by z amount. While the rectangular coordinates are the most common, some problems are easier to analyze in alternate coordinate systems. Other common systems are the 2D Polar coordinate system (Fig. 11.5b), 3D cylindrical coordinates (Fig. 11.5a), and spherical coordinate system (Fig. 11.6a). This next guided inquiry explores the definitions of different coordinate systems and how they relate to the rectangular system.

11.3.22 In Fig. 11.5b, what is the shape that is shown from the dashed lines connecting the point A to the x-axis?

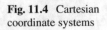
Fig. 11.4 Cartesian coordinate systems

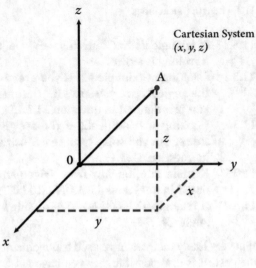

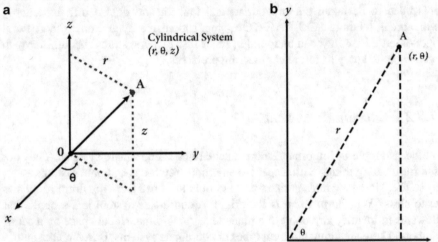

Fig. 11.5 (a) Image of cylindrical coordinate systems, then (b) 2D x–y plane of cylindrical system, which is known as the Polar Coordinate system

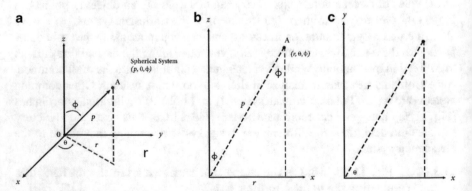

Fig. 11.6 (a) Image of spherical coordinate system, then (b) x–z plane, and (c) x–y plane on the system

11.3.23 Develop a relationship between dotted line *r* and the *x*- and *y*-axis.

11.3.24 Using the answer to the previous question, develop a relationship between *x*-axis line and the dashed line labeled *r*. (Hint: look back at Fig. 11.3a)

11.3.25 Using the answer to the previous two questions, develop a relationship between *y*-axis line and the dashed line labeled *r*. (Hint: look back at Fig. 11.3a)

11.3.26 Looking at Figs. 11.4 and 11.5a, what is the relationship between the *z*-axis of the Cartesian system and the *z*-axis of the Cylindrical system?

11.3.27 Using the answers to questions 11.3.22 through 11.3.26, explain the relationship between Polar coordinates and Cylindrical coordinates.

11.3.28 Using the answers to questions 11.3.22 through 11.3.27, write out an equation for *x*, *y*, and *z* that would convert between Cartesian coordinates and Cylindrical coordinates.

11.3.29 Convert the point (1, 1, 1) in rectangular coordinates to cylindrical coordinates.

11.3.30 Figure 11.6b shows the 2D projection of the *x–z* plane for spherical coordinates. What is the shape that is shown from the dashed lines connected the point *A* to the *x*-axis?

11.3.31 Develop a 3D relationship between dotted line ρ and the *x*, *y*, and *z*-axes.

11.3.32 Comparing Fig. 11.6a and b, what is the *x*-axis distance to point *A*?

11.3.33 Using the answer to the previous question, develop a relationship between *x*-axis line and the dashed line labeled ρ. (Hint: look back at Fig. 11.3a)

11.3.34 Comparing Fig. 11.6a and b, what is the *z*-axis distance to point *A*?

11.3.35 Using the answer to the previous question, develop a relationship between *z*-axis line and the dashed line labeled ρ. (Hint: look back at Fig. 11.3a)

11.3.36 Now, Fig. 11.6c shows 2D projection of *x–y* plane. What is the shape that is shown from the dashed lines connected the point to the *x*-axis?

11.3.37 In Fig. 11.6c, what is the length diagonal dotted line?

11.3.38 Using the same procedure as before (right angle equations), develop a relationship between *x*-axis line and the dashed line labeled *r*.

11.3.39 Thinking back to question 11.3.33, what relationship was developed for the diagonal *r* in terms of angle φ?

11.3.40 Using the same procedure as before (right angle equations), develop a relationship between *x*-axis line and the dashed line labeled *r* in terms of the angles φ and θ.

11.3.41 Using the same procedure as in questions 11.3.39 and 11.3.40, develop a relationship between *y*-axis line and the dashed line labeled *r* in terms of the angles φ and θ.

11.3.42 Use the answers to questions 11.3.31, 11.3.35, 11.3.40, and 11.3.41 to define the relations between spherical coordinates and rectangular coordinates.

11.3.43 Convert the point (1, 1, 1) in rectangular coordinates to spherical coordinates.

11.4 Important Engineering Parameters of Length

11.4.1 Area

Recall from Chap. 10 that Area is a secondary physical quantity, and it plays an important role in many engineering applications. The next set of questions explores the importance of the parameter area and discovers its definition.

11.4.1 **You have a glass of crushed ice and a glass of ice cubes, as shown in Fig. 11.7. When placed out in the sun, which glass of ice would melt faster? Explain why.**

11.4.2 **Figure 11.8a shows a figure of a desk, Fig. 11.8b shows the shape of the book on the desk if we look directly down at the desk. What is the shape of the book? What is the shape of the top of the desk?**

11.4.3 **In Fig. 11.8, how many books does it take to completely cover the desk?**

11.4.4 **Based on the previous questions, in your own words, what is the definition of *area*?**

11.4.5 **How do you do determine an area?**

11.4.6 **Name the fundamental dimension of area.**

11.4.7 **Name the fundamental units of area in SI units and US units.**

11.4.8 **Give three examples of where area plays an important role in engineering.**

There are many ways to calculate the area of something. The areas of common shapes, such as those given in Table 11.3 are known as *primitive areas*. However, typically engineers work with much more complex areas. These are known as *composite areas*. Engineers calculate composite areas by dividing them up into primitive areas. For example, the surface of an I-beam, such as shown in Fig. 11.9, can be determined from the sum of the primitive areas of different sized rectangles.

There are the other ways to calculate Area including the following:

Fig. 11.7 (**a**) Glass of crushed ice, (**b**) glass of cubed ice

(a)   (b)

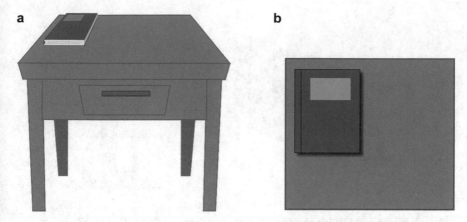

Fig. 11.8 (a) Desk with book and (b) top view of the top of book on desk

Table 11.3 Some Common Primitive Area Formulas and surface area formulas

Shape	Area Formula	Shape	Area Formula
Triangle		Rectangle	
Trapezoid		Circle	
n-Sided Polygon		Parallelogram	

Shape	Surface Area Formula	Shape	Surface Area Formula
Sphere		Cylinder	
Right Circular Cone		Ellipse	

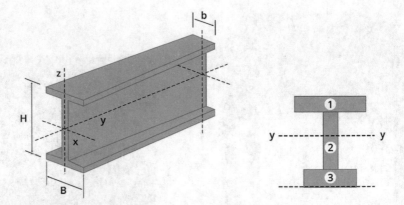

Fig. 11.9 I-beam with composite area made up of primitive area rectangles 1, 2, and 3

- **Trapezoidal rule**—Approximate the planar areas of an irregular shape by dividing the total area into small trapezoids of equal height, then summing up the trapezoids.
- **Counting the Squares**—Begin by dividing the given area into small squares of a known size, and then count the number. The not full squares at the end can be approximated as triangles.
- **Subtracting Unwanted Areas**—First take the area of a larger primitive area that is easy, then subtract the smaller unwanted areas.
- **Weighing the Area**—This procedure requires the use of an analytical balance and is best used for very complicated shapes. Start by weighing a known planar area (E.g., 8½" × 11" sheet of paper) and calculate the area. Then cut the unusual shape out of the same material that was just weighed, for example, cut the unusual shape out of a piece of paper. Then compare the ratio of the area of square paper over the weight of the square paper to the ratio of the area of the shape over the weight of the shape. Finally, solve for the area of the shape. This is approximate (not perfect, but close) as long as we assume uniform thickness and density of the object (in this case the paper).

The next set of questions will allow you to practice calculating area using the various methods.

11.4.9 **Figure 11.10 shows a picture of a composite shape. Explain why it is a composite shape.**

11.4.10 **In Fig. 11.10, what are the primitive areas used to make the composite shape?**

11.4.11 **What is the area of the solid shape?**

11.4.12 **What is the area of the primitive shape that is missing?**

11.4.13 **Using the previous questions, develop a relationship for the composite shape. Develop the equation for the hole size (D) in terms of shaded area (A) and , side S.**

Fig. 11.10 Composite
Figure of an equilateral
plate (of side length S). It
has a circular cutout (of
diameter D) at its center

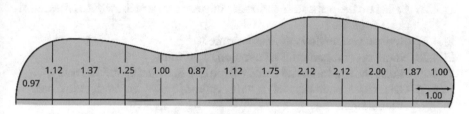

Fig. 11.11 Picture of the area that a groundskeeper needs to mow for questions 11.4. All of the
numbers are in units of feet

11.4.14 **Figure 11.11 shows a picture of an odd shaped lawn that needs to be
mowed. To calculate the area of the lawn using the trapezoidal rule,
how many trapezoids has the area been split into?**

11.4.15 **Looking at Fig. 11.11, which dimension of the trapezoid is constant.
The width or the height?**

11.4.16 **What is the width of each of the trapezoids?**

11.4.17 **What is the equation for the area of a trapezoid?**

11.4.18 **Looking at Fig. 11.11, for the first trapezoid on the left, what are the
measurements for the values of each of the variables in the area of a
trapezoid equation from the previous question.**

11.4.19 **Looking at Fig. 11.11, and using the previous answer, what is the area
for the first trapezoid on the left?**

11.4.20 **Looking at Fig. 11.11, for the second trapezoid from the left, what are
the measurements for the values of each of the variables in the area of
a trapezoid equation from question 11.4.17.**

11.4.21 **Looking at Fig. 11.11, and using the previous answer, what is the area
for the second trapezoid from the left?**

11.4.22 **Using the trapezoid rule, calculate an approximate area for the lawn in
Fig. 11.11.**

11.4.23 **Using your own methods (your memory, book, or internet search) fill
in Table 11.3 with the area and surface area formulas for the various shapes.**

11.4.2 Volume

Volume is another secondary physical quantity engineers use regularly. The next set of questions explores the importance of the parameter volume and discovers its definition.

11.4.24 **If you have two cups filled with coffee, shown in Fig. 11.12. Do the cups have the same amount of coffee in them?**

11.4.25 **Which cup can hold the most coffee in it? Explain why?**

11.4.26 **Based on the previous questions, in your own words, what is the definition of** *volume*?

11.4.27 **How do you do determine a volume?**

11.4.28 **Name the fundamental dimension of volume.**

11.4.29 **Name the fundamental units of volume in SI units and US units.**

11.4.30 **Give three examples of where volume plays an important role in engineering.**

The SI base unit of volume is the liter (L). To convert between length units and volume units, you can think about a cube filled with water. A cube with sides of length 10 cm each can hold 1 L of water ($1\ L = 10\ cm^3$). Another common SI volume unit is the milliliter (mL). A cube with sides of a length of 1 cm each can hold 1 mL of water ($1\ cm^3 = 1\ mL$). The US base unit of volume is the ounce. Originally derived from the Avoirdupois measurement system [2], it is defined as 1/128 of a US gallon. Other US volume units include the teaspoon, tablespoon, cup, pint, quart, gallon, and barrel. To convert between length units and volume units, 1 cubic inch (in.^{3}) is 0.55 fluid ounces. The same techniques used to calculate area can be applied to calculate the volume of an object, namely common volumes and composite volumes. Common volume shapes are given in Table 11.4. The next set of questions will allow you to practice calculating volume.

11.4.31 **Figure 11.13 shows ice cream scooped into two different shaped cones. What volumes can be used to represent each cone?**

11.4.32 **What is the volume of the ice cream held by the cake cone shown in Fig. 11.13a? (Note: include the scoop of ice cream on the top)**

11.4.33 **What is the volume of the ice cream held by the sugar cone shown in Fig. 11.13b? (Note: include the scoop of ice cream on the top)**

Fig. 11.12 Picture of two cups with different volumes

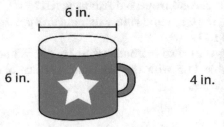

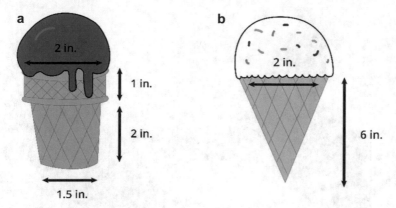

Fig. 11.13 Picture of ice cream held in (**a**) a cake cone and (**b**) a sugar cone

Table 11.4 Some common primitive volume formulas

Shape	Volume	Shape	Volume
Sphere		Cylinder	
Right Circular Cone		Section of a Cone	

11.4.34 **Based on the previous questions, explain which cone will hold the most ice cream.**

11.4.35 **Using your own methods (your memory, book, or Internet search) fill in Table 11.4 with the area and surface area formulas for the various shapes.**

11.4.3   Second Moment of Area

The second moment of area, also known as the area moment of inertia, or second area moment, is a length parameter which reflects how the geometry of an object is distributed with regard to an axis. In engineering, it is a shape property that significantly affects deflection and stress in beams. The symbols used to represent the

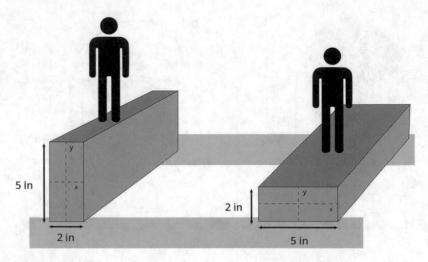

Fig. 11.14 (a) People standing on plans of different cross-sectional areas

second moment of area is typically an I for an axis that lies in the plane or a J for an axis perpendicular to the plane. In general, you can find the second moment of area by taking the integral of the square of the distance (r) times the area of the shape such that $I = \int r^2 A$. However, many common shapes have fully derived area moment of inertia formulas. The next set of questions works to discern the meaning to the second moment of area.

11.4.36 **In Fig. 11.14, a person stands on two beams, what is the shape of beam 1?**

11.4.37 **In Fig. 11.14, a person stands on two beams, what is the shape of beam 2?**

11.4.38 **In Fig. 11.14, what is the cross-sectional area of beam 1?**

11.4.39 **In Fig. 11.14, what is the cross-sectional area of beam 2?**

11.4.40 **For a rectangular cross section, the second moment of area $I = \dfrac{xy^3}{12}$, where x is the horizontal dimension of cross-sectional area, and y is the vertical dimension. Calculate the second moment of area for beam 1 in Fig. 11.14.**

11.4.41 **Calculate the second moment of area for beam 2 in Fig. 11.14.**

11.4.42 **In Fig. 11.14, which beam has a high second moment of area?**

11.4.43 **In Fig. 11.14, which beam bends the least?**

11.4.44 **In Fig. 11.14, which beam has a low second moment of area?**

11.4.45 **In Fig. 11.14, which beam bends the most?**

11.4.46 **Based on the previous questions, what information does the area moment of inertia provide?**

11.4.47 **For an I-beam like the one in Fig. 11.9, based on its cross-sectional area shape, explain the best way to get a high second moment of area?**

11.4.48 **What are the fundamental dimensions, SI units, and US units for the second moment of area?**

End of Chapter Questions

IBL Questions

IBL1: Based on Fig. 11.15 below, answer the following questions.

1. Consider the 1 m × 1 m × 1 m cube in Fig. 11.15a. What is the volume of the cube?
2. What is the area of one side of the big cube?
3. How many sides are there in Fig. 11.15a cube?
4. Based on the previous two questions, what is the exposed surface area of the cube?
5. If we divide each direction of the cube in half, we get eight smaller cubes as shown in Fig. 11.15b, what are the dimensions of one of those smaller cubes?
6. What is the total volume of all the smaller cubes?
7. What is the surface area of one small cube?
8. Based on the previous questions, what is the total exposed surface area of the smaller cubes?
9. If we divide each small cube into even smaller cubes, we get 64 smaller cubes with dimensions 0.25 m × 0.25 m × 0.25 m. What is the total volume of the smallest cubes and the total exposed surface area of the smallest cubes?
10. Describe the trend that is seen in the volume and surface area of the cube as it gets cut into smaller pieces.

IBL2: Based on Fig. 11.16, answer the following questions. Please note, for a symmetrical shape as shown in Fig. 11.16, the area moment of inertia around each axis can be calculated by

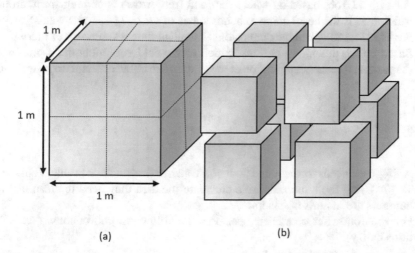

1 m

1 m

1 m

(a)

(b)

Fig. 11.15 Area and volume relationship between (**a**) One large cube and (**b**) smaller cubes

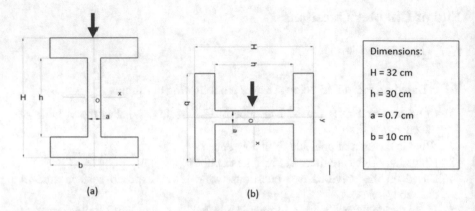

Fig. 11.16 Red arrow indicated the direction of loading for and (**a**) I-beam configuration and a (**b**) H-beam configuration

$$I_x = \frac{ah^3}{12} + \frac{b}{12}\left(H^3 - h^3\right)$$

$$I_y = \frac{a^3 h}{12} + \frac{b^3}{12}\left(H - h\right)$$

1. For Fig. 11.16a, based on where the load (red arrow) is coming from, around which axis is the beam going to bend; x-axis or y-axis?
2. For Fig. 11.16a, given the dimensions, calculate the area moment of inertia.
3. For Fig. 11.16b, based on where the load (red arrow) is coming from, around which axis is the beam going to bend; x-axis or y-axis? (Answer: y-axis)
4. For Fig. 11.16b, given the dimensions, calculate the area moment of inertia.
5. Based on the previous questions, which way would you orient the beam, in the shape of an I-beam configuration or the shape of an H-beam? Explain your choice.

Practice Problems

1. A lifeguard needs to vacuum a pool at a water park with an area shaped like Fig. 11.17. Using the trapezoidal rule calculate the area they need to clean. All the numbers are in units of meters.
2. For the following coordinate, perform the following conversions. (see solutions below)

   (a) Convert (2, 2, 4) from Cartesian to Cylindrical Coordinates
   (b) Convert (3, π/2, -5) from Cylindrical to Cartesian Coordinates

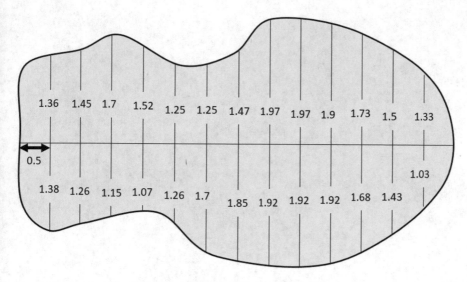

Fig. 11.17 Pool area needing to be vacuumed divided into trapezoids

    (c) Convert (, 4, -6) from Cartesian to Spherical Coordinates
    (d) Convert (4, π/4, π/6) from Spherical to Cartesian Coordinates

3. A metal pipe needs to be fabricated that is 150 ft long. The inner and outer diameter of the tube need to be in. and in., respectively. How much volume of material is needed to make the pipe?
4. Based on Fig. 11.2a develop a relationship between the angle units radians and the angle units degrees.

References

1. Trigonometry, R. Nagel, Encyclopedia of Science, 2nd Ed., The Gale Group (2002).
2. Chisholm, Hugh, ed. (1911). "Avoirdupois". Encyclopædia Britannica. 3 (11th ed.). Cambridge University Press. p. 66.

Chapter 12
Time

Abstract On the journey to become an engineer you will learn to appreciate and recognize the role of time and its related parameters. Specifically in this chapter, you will consider the parameters of frequency and period, which you will utilize in the future analysis of dynamics of mechanical systems and vibrations. You will learn to recognize the conditions of steady and unsteady states, which you will use in future heat transfer and fluid mechanics calculations. You will then comprehend the concepts of linear versus angular motion, and the associated velocities and accelerations. These concepts will be used in future dynamics, solid mechanics, and machine design classes.
By the end of this chapter, students will learn to:

- Explain the importance of the notion of time in engineering analysis.
- List examples of time and related variables.
- Explain the base units of time.
- Explain when the units of time are scaled appropriately and understandably.
- Define steady and unsteady state systems.
- Compare steady and unsteady state systems.
- Identify steady and unsteady state systems.
- Define period and frequency.
- Explain the relationship between period and frequency.
- Identify periodic examples in life and in engineering.
- Describe engineering quantities that are based on the fundamental dimensions of length and time.
- Define velocity, acceleration, speed (both average and instantaneous), and volume flow rate.
- Calculate speed, acceleration, and volume flow rate given certain engineering situations.
- Explain the relationship between linear speed and angular speed.

12.1   Importance of Time Parameters

Early on, humankind realized that they live in a dynamic world. Everything, including the universe, is constantly moving. Time became the means by which humanity measures its existence. Time is often referred to as a fourth dimension, along with three spatial dimensions [1]. This inquiry explores the importance of time and identifies its many effects on our lives and engineering factors.

12.1.1. **List three examples of questions frequently asked in our daily lives dealing with time.**
12.1.2. **Based on the previous question, in your own words how would you define time?**
12.1.3. **Based on the previous question, in your own words explain why is time important?**
12.1.4. **Give some examples of the role time plays in technology and engineering.**

12.2   Measurement of Time

Methods of progressive measurement take two distinct forms: the calendar and the clock [2]. A calendar is a mathematical tool for organizing intervals of time that are longer than a day. The clock is a physical mechanism that counts the passage of time for periods of less than a day. Increasingly however personal electronic devices display both calendars and clocks simultaneously. The number (as on a clock dial or calendar) that marks the occurrence of a specified event as to hour or date is obtained by counting from a central reference point.

Time standards were historically based on the rotation of the Earth because from the late eighteenth to nineteenth centuries it was assumed that the Earth's daily rotational rate was constant. In the early twentieth century, once it was confirmed that the earth's rotation is irregular, time standards became based on earth's orbital period and the motion of the Moon.

Then, in 1955, the invention of the cesium atomic clock led to the replacing astronomical time standards by newer time standards based on atomic time [3]. The second is defined as the interval of 9,192,631,770 periods of radiation corresponding to the transition between the two hyperfine levels of the ground state of a cesium 133 atom [4]. It is important to remember that when measuring time, it needs to be expressed in a way that (1) is scaled appropriately for the situation and (2) people can understand. The next inquiry explores the units and the expression of time.

12.2.1. **What is the base SI Unit for time?**
12.2.2. **What is the base UN Unit for time?**
12.2.3. **Explain the difference between SI and US Units for time?**

12.2.4. **If a person said that, they were 599,184,000 sec old on their birthday, how old are they so that people could understand.**

12.2.5. **List four other examples of time divisions (other than seconds) that are used to scale the dimension time.**

12.3   Steady Versus Unsteady State

The role of time is prevalent in engineering problems and solutions. Specifically, lots of engineering problems that deal with time are divided into two areas: *steady* state and *unsteady* state problems. Engineering examples of steady and unsteady state processes include cooling electronics equipment, combustion, materials casting, plastic forming, heating or cooling a building, and food processing. This next inquiry helps us to define, compare, and identify steady and unsteady state situations.

12.3.1. **Figure 12.1 shows milk being poured out of a glass container. Over time as the milk is poured out, what happens to the size and shape of the glass container?**

12.3.2. **Based on the previous question, would you consider the size of the container over time steady or unsteady?**

12.3.3. **Figure 12.1 shows milk being poured out of a glass container. Over time as the milk is poured out, what happens to the amount of milk in the container?**

12.3.4. **Based on the previous question, would you consider the amount of milk in the container over time steady or unsteady?**

12.3.5. **Based on the previous questions, define *steady* state.**

12.3.6. **Based on the previous questions, define *unsteady* state.**

12.3.7. **Look at the table of parameters in Table 12.1. List whether they are steady or unsteady with respect to time. List two more examples of each state (steady and unsteady).**

Fig. 12.1 Milk pouring
out of a glass container

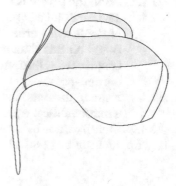

Table 12.1 Parameters that may change with time

Parameter	What is the parameter? Does it change with time?	Steady or unsteady state?
A person's shoe size		
Weight of a textbook		
A running sink with the drain open—where the amount of water entering the sink is less than the water draining out of the sink		
A running sink with the drain open—where the same amount of water entering the sink is draining out of the sink		
The velocity of a car being driven at a constant speed on the highway		
The position of a car being driven at a constant speed on the highway		

12.4   Period and Frequency

Many phenomena in life repeat themselves in regular intervals. These events or processes are known as cyclical, or periodic. To quantify a cyclical event two terms are typically used: period and frequency. Period is typically denoted by the capital letter T. Frequency is denoted by the lowercase letter f. Period is expressed in units of seconds per cycle and frequency is expressed in cycles per second, also denoted as Hertz (Hz). This next inquiry will help us define and identify the time parameters period and frequency and explain the relationship between them.

12.4.1.   **When is your birthday?**

12.4.2.   **How *often* does your birthday occur?**

12.4.3.   **What is the frequency of your birthday?**

12.4.4.   **How long does it take for your birthday to *reoccur*?**

12.4.5.   **What is the period of your birthday?**

12.4.6.   **From the previous questions, define period and frequency.**

12.4.7.   **The earth takes 1 year to travel around the sun; therefore, the period of the Earth's orbit is 1 year and its frequency is one orbit per year. Based on this statement explain the relationship between period and frequency. {Hint: Also check the units of f and T in the above paragraph}**

12.4.8.   **Explain the relationship between period and frequency in the form of a mathematical equation.**

12.4.9.   **Identify three other periodic examples in life.**

12.4.10.   **Identify three other periodic examples in engineering.**

12.5   Length and Time Engineering Variables

Even though time is a primary dimension, many secondary dimensions in engineering involve time and another fundamental dimension. Particularly many derived dimensions are based on time and length, such as speed, acceleration, and volume flow rate. The next inquiry helps us to look more in depth at each parameter.

12.5.1. **If a person travels a short distance in a long amount of time, are they considered fast or slow?**

12.5.2. **If a person travels a short distance in a long amount of time, do they have a high velocity or a low velocity?**

12.5.3. **If a person travels a large distance in a short amount of time, are they considered fast or slow?**

12.5.4. **If a person travels a large distance in a short amount of time, do they have a high velocity or a low velocity?**

12.5.5. **Based on the previous questions, define velocity.**

12.5.6. **What are the fundamental dimensions of velocity?**

12.5.7. **What are the SI and US units of velocity?**

12.5.8. **Name two other velocity units (one is SI and the other is US units) that are used for fast-moving objects. {Hint: Think about driving on a highway in the United States and then in Europe}.**

As explored in the previous inquiry, *Linear Velocity* is the time rate of change of position of an object moving along a straight path. It is a vector quantity, meaning that it describes both magnitude and direction. The magnitude component of velocity is called speed. *Average speed* is found by taking the ratio of the distance traveled over time. *Instantaneous speed* is the actual speed at any given instant of time. The next inquiry helps us distinguish between average speed and instantaneous speed. Read the following scenarios:

Scenario 1: You are traveling from New York City to Washington D.C., which is around 224 miles. The total trip takes you 4 h. You travel for an hour and a half and go 79.5 miles and then stop for food. Then you hit traffic due to construction and travel only 15 miles for 30 min. Then you travel the rest of the way (130 miles) for the rest of the trip.

12.5.9. **Given the above scenario, calculate the speed traveled during the first leg of the trip.**

12.5.10. **Given the above scenario, calculate the speed traveled when the car was in traffic.**

12.5.11. **Given the above scenario, calculate the speed traveled for the last leg of the trip after traffic.**

12.5.12. **Was the car traveling at the same speed throughout the entire trip?**

12.5.13. **Calculate the average speed for the entire trip.**

12.5.14. **Based on the previous questions, how would you define average speed?**

12.5.15. **Based on the previous questions, how would you define instantaneous speed?**

12.5.16. **For the above scenario, do we know the precious location of the car or the direction it is traveling?**

12.5.17. **For the above scenario, can we calculate the velocity of the car?**

12.5.18. **Explain what is the difference between a scalar and a vector quantity.**

12.5.19. **Give three examples of other engineering scalar quantities.**

Another time/length parameter is acceleration. In day-to-day life, the word acceleration is typically used to describe when something increases speed. However, in engineering, acceleration is more than just that. *Linear Acceleration* is the rate at which the linear velocity of an object changes over time. Acceleration has the same properties as velocity, namely it is a vector. Therefore, the velocity of an object can change over time when it is speeding up (typically called acceleration), slowing down (typically called deceleration), or changing direction. Even if the speed of an object does not change, changing the direction of motion changes the acceleration because it changes the vector. One of the most important accelerations that an engineer encounters is the *acceleration due to gravity*. This force attracts objects to the earth. The acceleration due to gravity is symbolically represented by the letter g, and it is a constant value of 9.81 m/s^2 or 32.2 ft/s^2, depending on which unit system the calculations used. *The next inquiry explores the parameter linear acceleration.*

   Scenario 2: Initially a car is driving 10 miles per hour down a street. Then the car speeds up to 50 miles per hour. The increase in speed takes 5 s.

12.5.20. **How long did it take the car to speed up?**

12.5.21. **How much did the car's speed increase in that time?**

12.5.22. **What was the car's change in velocity?**

12.5.23. **How fast did the car's speed increase per second?**

12.5.24. **What was the car's acceleration?**

12.5.25. **Define and write a mathematical expression for average acceleration.**

12.5.26. **If a car is driving at a constant velocity, what is its acceleration?**

If the linear position of an object can be tracked over time, the velocity and acceleration can be found. This next inquiry gives us practice describing the full motions of an object moving in a straight line (position, velocity, and acceleration).

   Scenario 3: A student is riding their bike to class and tracking their position. Class is 5 miles away. Figure 12.2 a shows a picture of the bike's change in position. Each red dot is the position on the diagram separated by an equal time interval of 1 min.

12.5.27. **Looking at Fig. 12.2 a, how far did the student get in the first 4 min?**

12.5.28. **Looking at Fig. 12.2 a, how long did it take the student to get 4 miles?**

12.5.29. **Looking at Fig. 12.2 a, how long did it take the student to get to class?**

12.5.30. **Draw a representative graph of the student's position versus time (note: you can plot the time on the abscissa and position on the ordinate).**

12.5.31. **Using the answers to the previous questions, what is the velocity of the student biker?**

12.5.32. **Draw a representative graph of the student's velocity versus time (note: you can plot the time on the abscissa and velocity on the ordinate).**

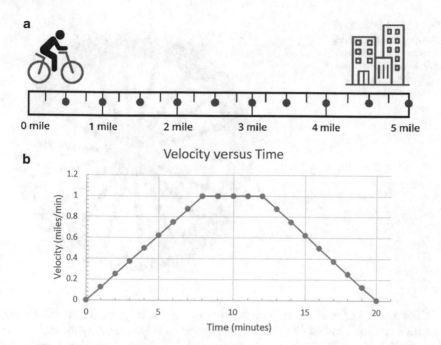

Fig. 12.2 (a) Showing a student biking to class, red dots indicate position for every 1 min recorded and (b) Graph of velocity versus time graph as the student bikes to their friend's dorm after class

12.5.33. **Using the answers to the previous questions, what is the acceleration of the student biker?**

12.5.34. **Figure 12.2 b shows the velocity of the student biking to a friend's dorm. Using Fig. 12.2 b, give the position, velocity, and acceleration of the student after 8 min have passed.**

12.5.35. **Using Fig. 12.2 b, give the position, velocity, and acceleration of the student between 8 and 12 min into the trip.**

12.5.36. **Using Fig. 12.2 b, give the position, velocity, and acceleration of the student between 12 min and the conclusion of the trip.**

12.5.37. **Using the previous questions, how far is the dorm from where the student left?**

12.5.38. **What is the average velocity for the trip?**

Just like body can move in a straight line and have linear position velocity and acceleration, a body can also rotate. Understanding rotational motion is also very important for engineers. Examples of engineering components that rotate include wheels, gears, shafts, blades, screws, bearings, and fans. Just as with linear motion, engineers want to track rotational position, velocity, and acceleration.

When things spin or rotate about a center point, like the bike tire shown in Fig. 12.3, each point along the spokes follows a circular arc. The blue lines from the center of the tire to the edge is the radius of the tire or circle (r). The red curve is the distance traveled along the circular path from point A to point B. It is known as the arc length (s).

Fig. 12.3 A bicycle tire
showing the relationship
between arc length and
radial distance to define
angular position

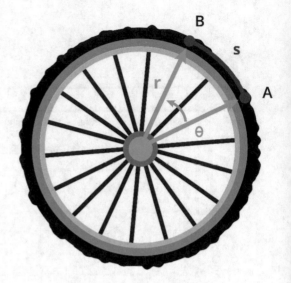

The amount of rotation that is occurring in the tire is quantified by the angular
position or displacement (θ). This angle of rotation is measured by the ratio of the
arc length to the radius of the circle. Namely, $\theta = \dfrac{s}{r}$. Angular position is analogous
to linear distance. The unit of angular position is the radian. The next set of questions helps us define the unit radian and see how it relates to our standard angle
measurement of degrees.

12.5.39. **We know that for one complete revolution of the bike tire, every point
on the tire is back to its original position, and thus the arc length is the
circumference of the circle. What is the equation for the circumference of a circle?**

12.5.40. **Based on the previous question, calculate one complete angular displacement for one revolution of the bike tire.**

12.5.41. **Based on the previous question, calculate how many radians are in
one circle.**

12.5.42. **Define the relationship between revolutions of a point along a circular
path and radians?**

12.5.43. **How many degrees are in one revolution or one full circle?**

12.5.44. **Based on questions 12.5.42 and 12.5.43, how many degrees are
in 1 rad?**

Comparable to linear velocity is angular velocity. The angular velocity of a point on
a rotating object is defined as the ratio of change in angular position (θ) over the
change in time (t) that it took the point to go through the angular displacement.
Angular velocity is typically denoted as the Greek letter omega namely $\omega = \dfrac{\theta}{t}$. It is
standard practice to express angular velocity in revolutions per minute (rpm);

however, the workable units (aka the units used in other equations involving angular velocity) are radians per second (rad/s), so you need to be familiar with converting between the two sets of units. This next inquiry gives us practice doing just that.

12.5.45. **How many radians are in one revolution? (hint: look back to question 12.5.44).**

12.5.46. **If a DVD rotates at a speed of nominally 1500 rpm, explain a procedure for how to convert from rpm to rad/s.**

12.5.47. **Calculate how fast the DVD in the previous question is spinning in radians per second?**

12.5.48. **If the earth goes through one full revolution in 1 day, calculate the angular velocity in rad/s of the earth.**

12.5.49. **How many degrees per hour does the earth rotate in 1 day?**

Then, there exists a relationship between the linear and angular velocities of an object that rotate and translate. The next set of questions helps us define the relationship between linear velocity (v) and angular velocity (ω) and gives us practice applying the relationship to everyday situations.

12.5.50. **In Fig. 12.3, if we consider a rock stuck on the bike tire at point A and it rotates to point B in a certain time (t), what is the rock's angular position?**

12.5.51. **In Fig. 12.3, if we consider a rock stuck on the bike tire at point A and it rotates to point B in a certain time (t), what is the rock's angular velocity?**

12.5.52. **In Fig. 12.3, if we consider a rock stuck on the bike tire at point A and it rotates to point B, what is the distance the rock translates?**

12.5.53. **Based on the previous question, how would you represent the rock's linear velocity (v)?**

12.5.54. **Based on the previous questions, develop a relationship between the rock's linear velocity and angular velocity.**

12.5.55. **If the tire in Fig. 12.3 belongs to a bicycle that is moving at a rotational speed of 60 rpm and the bike tire is 24 in. in diameter, what is the linear speed of the wheel in miles/h?**

For angular acceleration, recall that it is also a vector quantity and is defined in terms of angular velocity. Angular acceleration is the time rate of change of angular velocity. For example, if an ice skater is rotating on one skate, and then pulls their arms closer to their body, they will spin faster. Their angular velocity is increasing. Angular acceleration equals the change in angular speed over time and the faster the angular speed changes, the greater the angular acceleration. It is usually denoted by the Greek letter alpha, namely $\alpha = \dfrac{\omega}{t}$. Also, like velocity, there is a relationship between the linear acceleration of an object and its angular acceleration. Linear acceleration is directly proportional to angular acceleration multiplied by the distance from the center of rotation, namely $a = \alpha r$. This next set of questions helps us explore the parameter of angular acceleration.

12.5.56. **What are the standard units of angular velocity (hint: look at the previous section)?**

12.5.57. **Based on the previous question, what are the standard units of angular acceleration?**

12.5.58. **A motor shaft goes from zero to an angular velocity of 1500 revolutions per minute in 5 s. Calculate the angular acceleration of the shaft in standard units (hint: look at the previous question for units).**

12.5.59. **If the shaft in the previous question is slowed down, causing an angular deceleration of −87.3 rad/s², how long does it take for the shaft to stop spinning?**

This final length-time parameter to consider is **volume flow rate** (Q). This parameter determines the *dimensional* amount of a material or a substance that flows through a pipe. The volume flow rate is simply defined by the volume (V) of a given substance that flows through some point per unit time (t), namely $Q = \dfrac{V}{t}$. Because most fluids travel through pipes, there is a direct relationship between the volume flow rate, average velocity (v) of the substance in the pipe, and the cross-sectional area (A) of the pipe, $Q = v \times A$. This next inquiry explores the definition and application of the concept of volume flow rate.

12.5.60. **Given the definition of volume flow rate above, name the SI and US units of volume flowrate.**

12.5.61. **The pipe in Fig. 12.4 has water flowing through it. What is the shape of the cross-sectional area of the pipe?**

12.5.62. **For the pipe in Fig. 12.4, what is the area of the first section the water is flowing through?**

12.5.63. **If water is flowing through the pipe at an average velocity of 5 ft/s. Calculate the volumetric flow rate of the water in the pipe.**

12.5.64. **Now the water passes to the second section of the pipe. What is the area of the second section the water is flowing through?**

12.5.65. **If the water is steadily flowing, then the volumetric flow rate is constant between section 1 and 2. Based on that, and the previous question, what is the average speed of the water when it flows through section 2 of the pipe?**

12.5.66. **Is the average speed of the water when it flows through section 2 of the pipe the same, faster, or slower than when it flows through section 1? Explain your answer.**

12.5.67. **Predict what would happen if section 2 of the pipe was 12 in diameter?**

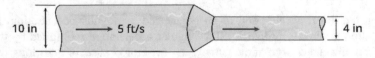

Fig. 12.4 Pipe system for volume flow rate inquiry

End of Chapter Questions

IBL Questions

IBL1: Using the picture below of a pendulum mass (m) in simple harmonic motion (SHM), answer the following questions:

1. Explain what is simple harmonic motion (SHM)?
2. Is the string that holds the mass tight or slack?
3. What two forces are on the mass?
4. The tightening force (P) felt in the string, what component of the weight of the mass is equal to T; $mgcos\theta$ or $mgsin\theta$?
5. The Resorting Force (R_F) is trying to get the mass back to equilibrium, what component of the weight of the mass is equal to R_F?
6. Newton's second law states that force is equal to mass times acceleration. Using Newton's second law and the restoring force, develop an equation for the acceleration of the pendulum mass.
7. What type of triangle is created by point ABC (green dotted lines)?
8. Create an equation for $sin\theta$ in terms of the length of the string (L) and the linear distance traveled by the mass (s).
9. Using questions 6 and 7, create an equation for the ratio of displacement/acceleration of the mass in terms of length of the string and distance traveled by the mass.
10. For SHM, the time period (T) is typically written as $2\pi\sqrt{\dfrac{\text{displacement}}{\text{acceleration}}}$.

    Using this definition and the previous question, write the equation for the motion of a pendulum in terms of the length of the string and acceleration due to gravity.
11. What is the length of a pendulum that has a period of 0.75 s?
12. What is the velocity of a pendulum when it reaches its maximum point during its swing?

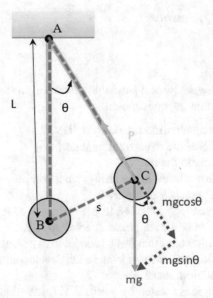

IBL2: The linear speed of a point on the surface of the earth varies with its location relative to the axis of the earth. At the equator, the earth's diameter is 7926 miles (12,756 km). Answer the following questions:

1. How long does it take for the earth to make one full revolution on its axis?
2. What is the relationship between one revolution and radians?
3. What equation can be used to relate linear velocity to angular velocity?
4. What is the linear speed of a point on the earth's equator due to the earth rotating in miles per hour?
5. What is the linear speed of a point on the earth's equator due to the earth rotating in kilometers per hour?

IBL3: A speed boat starts from rest and constantly accelerates to a speed of 50 mph in 20 s. Then the boat drives at a constant speed of 50 mph for the following 30 min. Then it takes the boat 30 s to constantly decelerate and come to a complete stop. Answer the following questions:

1. Draw a speed (mph) versus time diagram labeling the major values on the graph.
2. How does the velocity change during the first 20 s: linearly increases, linearly decreases, and remains the same?
3. Determine the average speed of the car during the initial 20 s.
4. Using the previous question, determine the distance traveled over the first 20 s.
5. Determine the distance traveled during the 30 min the boat was cruising at a constant speed.
6. How does the velocity change during the last 30 s: linearly increases, linearly decreases, and remains the same?
7. Determine the average speed of the car during the final 30 s.

8. Using the previous question, determine the distance traveled over the first 20 s
9. Using the previous question, determine the distance traveled over the final 30 s
10. Using the previous questions, determine the total distance traveled by the boat during this time.
11. Calculate the average speed of the boat for the entire trip.

Practice Problems

1. Determine the frequency of a pendulum whose string length is 20 feet.
2. A playground roundabout that is 10 ft in diameter is being spung by kids. If it takes 4 s for the kids to go from zero to a speed of 6 ft/s, calculate the final angular velocity and acceleration of the roundabout.
3. Using the data from IBL3, graph the acceleration of the boat as a function of time. Now plot each acceleration over time (put time in seconds to capture the shore timeframes
4. A water basin (dimensions 1.5 ft × 2 ft × 1.5 ft) is being filled by a hose with a diameter opening of 0.75 inches. If it takes 1 min to fill the basin up completely, calculate the volumetric flow rate and the average velocity of the water coming out of the faucet.

References

1. *About Time: Einstein's Unfinished Revolution*, Paul Davies, p. 31, Simon & Schuster, 1996, ISBN 978-0-684-81822-1.
2. Richards, E. G. (1998). Mapping Time: The Calendar and its History. Oxford University Press. pp. 3–5.]
3. P K Seidelmann & T Fukushima (1992), "Why new time scales?", Astronomy & Astrophysics vol.265 (1992), pages 833-838, including Fig. 1 at p.835, a graph giving an overview of the rate differences and offsets between various standard time scales, present and past, defined by the IAU.
4. Saeed Moaveni (2005). Engineering Fundamentals: An Introduction to Engineering (2nd ed.). Thompson. ISBN: 0-534-42459-7.

Chapter 13
Mass

Abstract Another important fundamental parameter in engineering is Mass. First, in this chapter, you will investigate mass by recalling matter, and its basic building blocs, the atom. Next, you will look at its applications to engineering, specifically material parameters such as density and specific gravity. Then you will learn about dynamic mass parameters such as mass moment of inertia, momentum and mass flow rate, and their engineering applications. Finally, you will investigate the mass-related concept of conservation of mass.
By the end of this chapter, students will learn to:

- Explain the importance of the notion of mass in engineering analysis.
- List examples of mass and related variables.
- Explain the base units of mass.
- Explain how the units of mass are scaled appropriately and understandably.
- Identify other mass-related quantities, and define their relationships with materials.
- Define and explore the mass moment of inertia parameter.
- Calculate the moment of inertia for uniformly shaped bodies.
- Define and compare linear momentum and angular momentum.
- Calculate linear and angular momentum in physical situations.
- Explain mass flow rate and its importance in engineering.
- Calculate mass flow rate of physical situations.
- Explain the principle of conservation of mass.
- Define a control volume in a mass balance problem.
- Write out a mass balance equation and solve for a specific situation.

13.1   Importance of Mass Parameters

The notion of quantity is a very old concept. Early humans realized that the number of objects in a group was directly proportional to the weight of the collection. They needed a term to quantify the physical phenomenon. Mass is a property of a

© Springer Nature Switzerland AG 2022
M. Blum, *An Inquiry-Based Introduction to Engineering*,
https://doi.org/10.1007/978-3-030-91471-4_13

physical body or something that has matter. Matter is defined as something that has mass and takes up space [1]. All objects and living things are made of matter, and matter is made of atoms or chemical elements. Recall from high school chemistry that atoms are the basic building blocks of all matter and are made of smaller particles called electrons, protons, and neutrons. Matter can exist in four states, depending on its surrounding conditions—solid, liquid, gaseous, or plasma state [2]. This inquiry helps us define and explore the importance of mass and has us identify how it effects our lives and its importance in engineering.

13.1.1. **Does a brick have mass?**

13.1.2. **Does a strawberry have mass?**

13.1.3. **Does smoke have mass?**

13.1.4. **List three other examples of things with mass in our daily lives.**

13.1.5. **If we have a piece of clay, and the clay gets thrown across the room, does the clay's mass increase, decrease, or remain the same?**

13.1.6. **Does mass change if a body moves or its position changes.**

13.1.7. **If we have a piece of clay, and we squash the clay, does the clay's mass increase, decrease, or remain the same?**

13.1.8. **If a body's shape is altered, but no material removed, does its mass change?**

13.1.9. **If we have a piece of clay, and we rip the clay in half and throw away one piece, does the clay's mass increase, decrease, or remain the same?**

13.1.10. **Does mass change if the material is removed or added to a body?**

13.1.11. **Based on the previous question, in your own words how would you define mass.**

13.1.12. **Based on the previous question, in your own words explain why the concept of mass is important.**

13.1.13. **Give examples of situations from everyday life where mass plays an important role.**

13.1.14. **Give examples of situations from engineering where mass plays an important role.**

13.2   Measurement of Mass Versus Weight

Early in history, humans did not fully realize the concept of gravity, so the distinction between mass and weight was initially unclear. Civilizations developed devices to measure amounts based on proportionality using balance scales [3]. The device, shown in Fig. 13.1, balances the force of one object's weight against the force of another object's weight. Since the measurements occurred on earth, the objects on each side of the scale experience similar gravitational fields. Hence, if they have similar masses then their weights will also be similar. This allows the balance scale to essentially compare masses, by comparing weights. In modern times, there are more sophisticated devices that can parse between weight and mass. Mass is also an inertia quantity. It describes an object's resistance to linear motion. If we think of

Fig. 13.1 Example of a balance scale measurement device

Newton's Second Law of Motion (Force = mass × acceleration), we see that the more mass an object has, the more force it will take to make it move.

The units of mass and mass conversions are sometimes confusing to students because each unit system, SI and US, use a different set of base units for mass. In the SI unit system, mass is a base unit, and then the unit of force is derived from Newton's second law. Therefore, in the SI system, the unit of mass does not come from any other form [4], and the secondary unit of force is based on the unit of mass. Originally, the SI unit was defined as the mass of 100 cm³ or 1 L of water, and today it is still very close to that amount. This next set of questions let us explore the SI unit of mass.

13.2.1. **List three verbal descriptions of mass. (They can be descriptive, do not need to be exact)**
13.2.2. **Force is an action applied to an object that causes it to accelerate, which is defined by Newton's second law. What is the name of the SI unit of Force (hint: what is the name of the second law? Also, check Fig. 10.3).**
13.2.3. **This defines the force required to accelerate a 1 kg object at a rate of 1 m/s². Based on the information and the previous question, what is the base units for the SI unit of Force?**
13.2.4. **In the definition of the SI unit of Force, what base units represent the acceleration of the object?**
13.2.5. **Based on the previous questions, what is SI unit of mass?**
13.2.6. **If a car had a mass of 700 kg on earth, what would the mass of the car be if the car was on the moon?**
13.2.7. **Is the SI unit for mass dependent on gravity?**

In the US unit system, unlike the SI system, the unit of mass is defined independently from the units of force. In the US system, the unit of force is the Pound Force, abbreviated lb$_f$, and the unit for mass is the Pound Mass, abbreviated lb$_m$. Often another mass unit is used called a *slug*. Various definitions have been used but the most common today defines the lb$_m$ as exactly 0.45359237 kg. Being able to distinguish between the different units and understanding how to convert correctly between them in calculations is very important in properly using US units in engineering. This next inquiry helps us understand the differences in the mass unit labels.

13.2.8. **What is Newton's second law?**

13.2.9. **What is Newton's second law written in fundamental dimensions if the unit for mass is not defined? (F = ?)**

13.2.10. **Using the previous question, what are the fundamental dimensions for mass using Newton's second law? (m = ?)**

13.2.11. **Using the previous question, what are the US units for mass based on Newton's second law?**

13.2.12. **If the mass unit slugs are derived from Newton's second law using US units, what is the definition of a slug?**

13.2.13. **What is the acceleration due to gravity in English units?**

13.2.14. **Based on the previous questions, what is the force required to accelerate an object with a mass of 1 lb_m at the acceleration due to gravity? (hint: Use Newton's second law)**

13.2.15. **Based on the previous question, what is the definition of a pound force?**

13.2.16. **Using questions 13.2.12 and 13.2.15, what is the relationship between pound mass and slug?**

Sometimes, another source of confusion for students is designation and conversation between weight and mass. Remember, the weight of an object is the force of a planet's gravity that is being exerted on that object. Mass differs from weight because, the mass of an object is always the same regardless of the object's location however the weight of an object can differ depending on where it is located and the gravity at that location. For the SI system, the weight unit is the same as the force unit. For the US unit system, when you measure weight using pounds, it should be pounds force that is being measured. Therefore, when you have an object and you are told its mass in lb_m, you can find its weight in lb_f, provided you know what the acceleration due to gravity is at that location, by using the following procedure:

- First, use $F = ma$ to find weight (in lb_m ft/s^2)
- Then convert the weight (lb_m ft/s^2) to lb_f using the relationship between lb_f and lb_m (1 lb_f = 32.2 lb_m ft/s^2)

This next inquiry gives us practice changing between them.

13.2.17. **A person goes bowling on earth. They use a bowling ball that weighs 10 lb_f. Write out a procedure for how to find what is the mass of the bowling ball in lb_m on earth?**

13.2.18. **Using the previously laid out procedure, what is the mass of the bowling ball in lb_m on earth?**

13.2.19. **An astronaut goes bowling on the moon, where the acceleration due to gravity is $a = 5.32$ ft/s^2. They measure the ball on the moon, and it weighs 10 lb_f. What is the mass of the bowling ball in lb_m on earth?**

13.2.20. **For the astronaut's bowling ball, what is the weight of the bowling ball on earth?**

13.2.21. **Is the bowling ball's lb_f measurement on earth the same as its lb_m measurement? Explain why or why not.**

13.2.22. **Does the bowling ball on earth have the same mass as the bowling ball moon?**

13.2.23. **Is the bowling ball's lb_f measurement on the moon the same as its lb_m measurement? Explain why or why not.**

13.3 Density, Specific Volume, and Specific Gravity

In engineering, many decisions are made concerning deciding the appropriate material to use for a specific application. Oftentimes, engineers are comparing the strength of a material to its mass/weight. Therefore, engineers define several terms to represent how light or how heavy materials are based on a unit volume. Listed are four mass-related material parameters.

- **Density** of a material is its mass per unit volume. It is typically symbolized by the Greek letter lower case ρ.
- **Specific volume** is the ratio of the material's volume to its mass. It is typically symbolized by the Greek letter lower case v.
- **Specific gravity** compares the ratio of the density of a substance to the density of a well-known standard. For a liquid or a solid, the standard comparison is the density of water at its densest at a temperature of 39.2 °F (4 °C). For a gas, the standard comparison is the density of air at room temperature of 68 °F (20 °C). It is typically symbolized by SG.
- **Specific Weight** measures the ratio of the weight of a material to the volume that it occupies. It is typically symbolized by the Greek letter lower case Υ.

This next set of questions lets us explore these mass-related material parameters.

13.3.1. **Which has more mass, one cubic foot of rubber or one cubic foot of aluminum?**

13.3.2. **Write out the definition of density in symbols (hint: recall the fundamental dimension symbols from Chap. 10).**

13.3.3. **Write out the US and SI Units for the definition of density (hint: recall the US units for mass).**

13.3.4. **Write out the definition of specific volume in symbols (hint: recall the fundamental dimension symbols from Chap. 10).**

13.3.5. **Write out the English and SI Units for the definition of specific volume.**

13.3.6. **From the previous questions, what is the relationship between density and specific volume?**

13.3.7. **The density of water at 4 °C is measured at 62.4 lb_m/ft^3. If liquid mercury has a density of 844.9 lb_m/ft^3, what is the specific gravity of liquid mercury?**

13.3.8. **Based on the previous question, what are the units for specific gravity?**

13.3.9. **Write out the definition of specific weight as a mathematical equation.**

13.3.10. **Based on Newton's second Law, how else can the term Weight be written?**

13.3.11. **What is the relationship between volume, mass, and density?**

13.3.12. **Based on the previous three questions, can you develop a relationship to relate specific weight and density of an object?**

13.4 Mass Moment of Inertia

Another mass quantity that is frequently used in mechanical design calculations is the **mass moment of inertia**. Recall how the property of mass characterizes an object's resistance change in linear motion. Similarly, mass moment of inertia represents an object's resistance to change in rotational motion. As an object is rotating about a fixed axis, you can calculate how difficult it would be to change that object's rotational speed about that axis. The moment of inertia is dependent on where the mass is distributed within the object and the position of the axis. This means that one object can have different moment of inertia values, based on where the axis of rotation is oriented relative to the body. The *mass moment of inertia, relative to a given axis of rotation, is calculated by multiplying the mass by the square of the distance from the axis of rotation. If there are multiple masses, the mass moment of inertia is found by summing up each product of mass and squared distance.* The variable for mass moment of inertia is typically designated with the letter capital I, just like area moment of inertia (think back to Chap. 10), but the units are different. The next set of questions works to explore the meaning of the mass moment of inertia.

13.4.1. **Looking at Fig. 13.2a, write an equation for the mass moment of inertia of the single mass m. (Hint: look at the italics portion of the previous text for the formal definition of I).**

13.4.2. **Looking at Fig. 13.2b, write an equation for the mass moment of inertia of the multiple mass system using all the masses. (Hint: look at the italics portion of the previous text for the formal definition of I).**

13.4.3. **What are the fundamental dimensions for mass moment of inertia?**

13.4.4. **What are the SI and US units for mass moment of inertia?**

13.4.5. **Recall what are the SI and US units for the area moment of inertia (hint, look at Chap. 11, question 11.4.46).**

13.4.6. **Figure 13.3 gives the mass moment of inertia equations for three common objects. If you were to race these objects down a hill as shown in the figure, and they all had the same radius and mass, explain which object would win the race.**

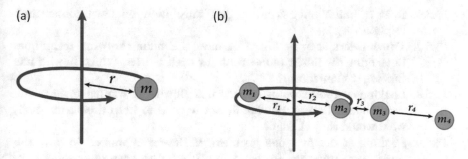

Fig. 13.2 Examples of mass moment of inertia for (**a**) a single mass (**b**) multiple masses

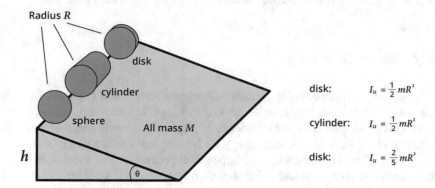

Fig. 13.3 Different shapes racing down a hill

13.5 Momentum

Mass is also an important consideration when objects are moving. If two balls were thrown at the same speed, a bowling ball and a football, and you had the option of catching one, which one would you catch? Probably the football, right? The reason you would opt to catch the football is because it would hurt less when it hit you because the football is lighter than the bowling ball. It has less mass even though both are moving at the same speed. That is what the variable **linear momentum** describes. It quantifies mass in motion. Linear momentum is defined as the product of mass and linear velocity. It is typically represented by the symbol $\vec{L}$. The next set of questions explores the concept of linear momentum.

13.5.1. **What is the symbol for the mass of an object?**
13.5.2. **What is the symbol for the linear velocity of an object?**
13.5.3. **Write an equation for the linear momentum of an object.**
13.5.4. **Explain why the symbol for linear momentum** $\vec{L}$ **has an arrow over it?**
13.5.5. **What are the SI and US Units for linear momentum?**

13.5.6. **Does a bullet shot from a gun have high or low momentum? Explain why.**

13.5.7. **Two bowlers seen in Fig. 13.4 have a similar throwing technique. Determine the linear momentum for each bowler. Which bowler has the larger momentum?**

13.5.8. **Looking at the previous problem. If Bowler A wanted to have the same momentum as Bowler B, but wanted to keep their same ball, what would they change?**

13.5.9. **Looking at the previous problem. If Bowler B wanted to have the same momentum as Bowler A, but wanted to keep their same ball, what would they change?**

13.5.10. **What is the momentum of an object at rest? Explain your answer.**

13.6   Mass Flow Rate

For engineers, it is important to identify and account for changes in physical quantities. Similar to volume flow rate (which was discussed in Chap. 12), one important quantity is **mass flow rate**. Mass flow rate is important in many engineering applications such as food production, energy for heating and cooling applications, and many other chemical, biological, and industrial processes. The symbol for mass flow rate is a lower case m with a dot above it ($\dot{m}$). Let us look at an example to help explain mass flow rate. Figure 13.5 shows two pictures (a) a filled up water balloon and (b) the water balloon when it has a small hole in it. Answer the following questions about the situation.

13.6.1. **In Fig. 13.5, what gives the balloon its size and weight?**
13.6.2. **What is the main mass of the water balloon?**
13.6.3. **Describe what is happening to the size of the water balloon in Fig. 13.5a to b.**

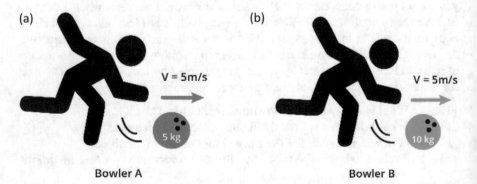

Fig. 13.4 Two bowling balls being thrown at velocity of 5 m/s with masses of (**a**) 5 kg and (**b**) 10 kg

Fig. 13.5 Water balloon   (a)   (b)
with a small hole in it
losing mass. (**a**) Before the
hole (**b**) after the hole

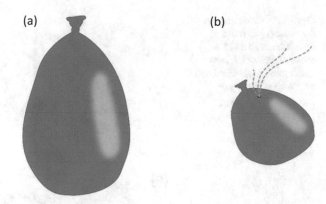

13.6.4. **Describe what is happening to the weight of the water balloon in Fig. 13.5a to b.**

13.6.5. **Why are the changes described in questions 13.6.31 and 13.6.4 occurring?**

13.6.6. **Based on your previous answers, what is the definition of mass flow rate?**

13.6.7. **Write a mathematical expression for mass flow rate.**

13.6.8. **What are the SI and US units for mass flow rate?**

Finding mass flow rate is important because it tells how much material is being added or removed from a process and is a very important part of the conservation of energy principle you will investigate in the next section. Mass flow rate can be found by measurement or calculation. To measure mass flow rate, two common methods are used. The first method measures mass flow rate by measuring the time it takes to reach a certain mass, shown in Fig. 13.6a. The second method measures how much mass accumulated over a specific amount of time, shown in Fig. 13.6b. However, these measurement methods are impractical for large-scale processes; therefore, engineers can calculate mass flow rate in two ways. The first is by measuring volume flow rate and then relating the two quantities. The second way is to measure the linear velocity of the substance and relate the two quantities. The next section will guide you through questions on how to calculate mass flow rate and give you practice calculating the mass flow rate of physical situations.

13.6.9. **Recall question 13.6.7, write the mathematical expression you developed for mass flow rate.**

13.6.10. **Is there a material property that relates the volume of a fluid to the mass of a fluid?**

13.6.11. **Write an equation for the relationship between mass and volume of a fluid.**

13.6.12. **Based on the previous two questions, develop a relationship between the mass flow rate of a substance and the volume flow rate of a substance.**

13.6.13. **Think back to Chap. 12, what is the relationship between volumetric flow rate and linear velocity of a fluid?**

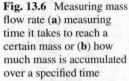

Fig. 13.6 Measuring mass flow rate (**a**) measuring time it takes to reach a certain mass or (**b**) how much mass is accumulated over a specified time

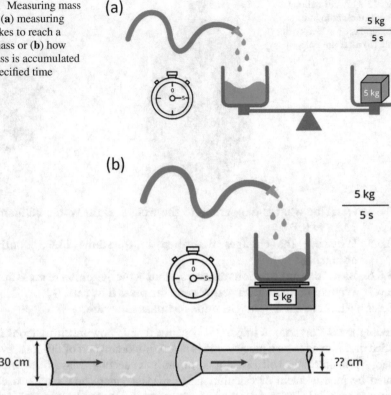

Fig. 13.7 Pipe system for mass flow rate inquiry

13.6.14. **Based on the previous question, develop a relationship between the mass flow rate of a substance and the linear velocity of a substance.**

13.6.15. **Scenario 1: The pipe in Fig. 13.7 has water (density is 1 g/cm³) flowing through its big section at 10 cm/s. What is the mass flow rate through the tube?**

13.6.16. **The mass flow rate of water through the little section of the pipe in Fig. 13.7 is 0.9 kg/s and the fluid velocity is 15 cm/s. What is the diameter of the little section of the pipe?**

As a final note, remember that the density and the volume of a substance can change, depending on where temperature and pressure are currently held. Therefore, it is important to remember to look up the density and/or volume of your substance using charts appropriate for the application.

13.7   Conservation of Mass

In the first section of this chapter, we defined mass as essentially the amount of matter a substance or object has. This next set of questions will help us define the **principle of mass conservation**. This principle is commonly used in many engineering applications, such as chemical reactions, mechanics, and fluid dynamics. Consider the following example:

Scenario: Mackenzie comes home from swimming practice and takes a shower in her bathtub. The tub has been recently cleaned and the drain is fully open as shown in Fig. 13.8a. Then her brother Tommy comes home from lacrosse practice and takes a shower. Grass from practice clogs the drain shown in Fig. 13.8b. They both run the shower so water comes out of the faucet on full blast.

13.7.1. **What happens to the water level in the tub while Mackenzie showers? Does it increase, decrease, or remain the same? Explain why.**

13.7.2. **What happens to the water level in the tub while Tommy showers? Does it increase, decrease or remain the same? Explain why.**

13.7.3. **For the two scenarios, how can we tell whether the tub will fill up or not?**

13.7.4. **In the two scenarios, what is the mass being transferred?**

13.7.5. **In the two scenarios, what is the space that the mass is being transferred through?**

13.7.6. **When the water is flowing through the tub, does the size of the tub change?**

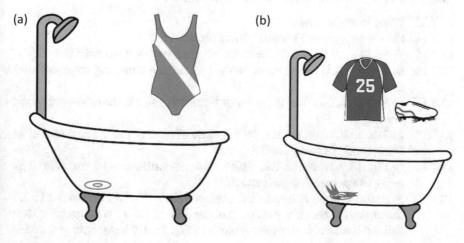

Fig. 13.8 Conservation of mass principle applied to a shower when emptying when the (**a**) drain is clear (**b**) the drain is clogged

13.7.7. **In the two scenarios, what could be considered the control volume (A control volume is a defined region of space that mass can move into and out of)?**

13.7.8. **Based on questions 13.7.5 through 13.7.7, what are two rules that should be set when making a control volume?**

13.7.9. **How could you use the idea of conserving mass through a constant space to describe the scenario given?**

13.7.10. **Based on the previous questions, explain in your own words the conservation of mass.**

13.7.11. **During Tommy's shower, the rate at which water is coming out of the faucet is 3 kg/s and the rate at which water is coming out the drain is 1 kg/s, how much water is in the tub after 10 min? If the tub is 1 m³ volume and the density of water is 1000 kg/m³, does it overflow?**

13.7.12. **After the next swim practice, Mackenzie wants to take a bath. She fills the tub at a rate of 1.5 kg/s. How long does it take her to fill the bathtub?**

For many engineering applications, particularly in fluid dynamics, the principle of conservation of mass developed above is adequate. However, there is more to consider than just mass flow in and out.

An important idea that the ancient Greek philosopher Empedocles first proposed was that matter never vanishes [5]. Have you ever heard the saying "You can't make something from nothing?" This is essentially the full version of the principle of conservation of mass, which states that *in a system, the amount of matter (mass) in the system can neither be created nor destroyed—it can only change form.* This next set of questions will let us explore this statement.

Scenario: Josiah is trying to defrost a large block of ice. He places it on a grate, in a tub with a drain, out in the sun. Also to also help the process he slowly pours warm water on it, as shown in Fig. 13.9.

13.7.13. **What is water made of?**

13.7.14. **How many physical states can water exist in?**

13.7.15. **What are the names of the states of matter water can exist in?**

13.7.16. **In Fig. 13.9, is the state of matter of the ice changing or remaining the same?**

13.7.17. **In Fig. 13.9, how many states of matter is the ice changing to? What are they?**

13.7.18. **In Fig. 13.9, from the ice, which state of matter can be considered as being made or *generated*?**

13.7.19. **In Fig. 13.9, from the ice, which state of matter can be considered as being evaporated or *consumed*?**

13.7.20. **A *system* can be defined as a control volume (recall questions 13.7.6 through 13.7.8), as a designated portion of the environment with a defined boundary. For the scenario in Fig. 13.9, what would you define as the system?**

13.7.21. **For the system just defined, name all the mass that is entering the system?**

(a)

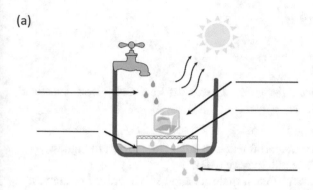

(b)

Conservation of Mass:

_____ + _____ - _____ - _____ = _____

Fig. 13.9 Defrosting a large block of ice. The ice is placed on a grate, in a tub with a drain, out in the sun. Also, warm water is dripping on it. (**a**) Figure (**b**) Equation

13.7.22. **For the system just defined, name all the mass that is leaving the system?**

13.7.23. **If the mass that is entering faster than the mass leaving the system what will happen?**

13.7.24. **Based on previous question, if mass is entering faster than the mass leaving the system, will mass accumulate or deplete?**

13.7.25. **Based on all the previous questions, how could you use the idea of conserving of mass through a system to describe the scenario given in Fig. 13.9.**

13.7.26. **Based on the previous questions, fill in the blanks in Fig. 13.9a with the following labels: mass-in, mass-out, accumulation, generation, and consumption.**

13.7.27. **Based on the previous question, in Fig. 13.9b complete the equation describing the conservation of mass principle with the following labels: mass-in, mass-out, accumulation, generation, and consumption.**

13.7.28. **If Josiah was melting a block of ice and used 2 kg of warm water from the hose, then 3 kg of liquid water was melted from the block, and, then 4 kg of water went through the drain, and since the drain was fully open no water accumulated in the tub, how much of the ice was being evaporated?**

Also, it is important to note, in reality, the mass of matter in an isolated system will always be constant regardless of any chemical reactions or physical changes that take place as long as it takes place in an *isolated* system. An *isolated* or *closed* system means that it does not interact with its environment (which our system did, because the sun was melting the ice).

End of Chapter Questions

IBL Questions

IBL1: For the following materials: gold, silver, Cast Iron, Oak Wood, and Sand, answer the following questions:

1. What is the density of water?
2. What is the density of each material listed in order from lowest to highest?
3. What is the definition of specific gravity (SG)?
4. Calculate the specific gravity of each material and list it in order from lowest to highest?
5. What is the relationship between density and specific gravity?
6. What is the value for specific gravity in SI units?

Practice Problems

1. Calculate the momentum for the following scenarios and explain which has the largest momentum.

   - A bullet with an average mass of 15 g moving at a speed of 800 m/s.
   - A lacrosse ball with a mass of 150 g moving at a speed of 45 m/s.
   - A baseball with a mass of 145 g moving at a speed of 50 m/s.

2. Determine the mass moment of inertia of a cylindrical rod made of stainless steel ($\rho = 7500$ kg/m^3) that is 1.5 m long and has a diameter of 7 cm.
3. A water basin (dimensions 1.5 ft $\times$ 2 ft $\times$ 1.5 ft) is being filled by a hose with diameter opening of 0.75 in. If it takes 1 min to fill the basin up completely, calculate the mass flow rate of the water coming out of the faucet.

References

1. *Matter (physics)*. McGraw-Hill's Access Science: Encyclopedia of Science and Technology Online. Archived from the original on 17 June 2011. Retrieved 2 February 2020.
2. M.A. Wahab (2005). Solid State Physics: Structure and Properties of Materials. Alpha Science. ISBN 978-1-84265-218-3.
3. L. Sanders, A Short History of Weighing, W. & T. Avery ltd, 2nd Edition, W. & T. Avery, 1960.
4. Robert L. Norton, Machine Design: An Integrated Approach, 5th Ed., Pearson, New Jersey, 2014. ISBN: 978-0-13-335671-7.
5. see pp.291–2 of Kirk, G. S.; J. E. Raven; Malcolm Schofield (1983). The Presocratic Philosophers (2nd ed.). Cambridge: Cambridge University Press. ISBN 978-0-521-27455-5.

Chapter 14
Force

Abstract Understanding how forces effect different objects is important in engineering. In fact, it is one of the fundamental disciplines of engineering, named Mechanics. There are three branches of mechanics; rigid body mechanics, deformable body mechanics, and fluid mechanics. An engineering student will begin to study rigid body mechanics typically in two classes; statics and dynamics. Then engineering mechanics fundamentals continue with the studies of strength of materials, failure of materials, and fluid dynamics. The principles of mechanics are used in many engineering disciplines, such as Civil, Mechanical, Aerospace, Environmental, and Biomedical engineering. The goal of this chapter is to introduce the main component of mechanics, namely the concept of force, its properties, its various types, and other force-related variables.

By the end of this chapter, students will learn to:

- Define what a force is.
- List and compare the different categories that forces fall under.
- Explain the effects that forces have on objects.
- Identify how to group forces when analyzing a physical situation.
- Define the fundamental dimensions of force.
- Define the SI and English units of force, and convert between them.
- Explain how to measure force.
- Identify the properties of force vectors.
- Resolve a force in two dimensions and combine a force into resultant components.
- Define, calculate, and apply different force parameters to engineering situations; including, pressure, stress, moment torque, and impulse.
- Explain why engineers use free body diagrams.
- Develop an effective strategy for analyzing a physical situation and identifying the forces that are present.
- Create an engineering drawing, aka free body diagram, of forces on rigid objects.

© Springer Nature Switzerland AG 2022
M. Blum, *An Inquiry-Based Introduction to Engineering*,
https://doi.org/10.1007/978-3-030-91471-4_14

14.1 Importance and Categories of Forces

Forces are responsible for so many physical phenomena and are so important in engineering. Understanding the concept of force allows engineers to be able to analyze many problems, design and manufacturing things like machines, robots, buildings, airplane and car frames, satellites, and medical implants. This next set of questions will help us to develop a comprehensive understanding of the physical phenomenon of force.

14.1.1. **A person is pushing against a brick wall as shown in Fig. 14.1a, what** *interaction* **is happening in this scenario?**

14.1.2. **A person is pushing against a brick wall as shown in Fig. 14.1a, what** *force* **is occurring in this scenario?**

14.1.3. **In Fig. 14.1a, what has the force resulted in?**

14.1.4. **Kids are making taffy as shown in Fig. 14.1c, what** *interaction* **is happening in this scenario?**

14.1.5. **Kids are making taffy as shown in Fig. 14.1c, what** *force* **is happening in this scenario?**

14.1.6. **In Fig. 14.1c, what has the force resulted in?**

14.1.7. **A jug falls off a table as shown in Fig. 14.1b, what** *interaction* **is happening in this scenario?**

14.1.8. **A cup falls off a table as shown in Fig. 14.1b, what force is happening in this scenario?**

14.1.9. **In Fig. 14.1b, what has the force resulted in?**

14.1.10. **Based on your answers to questions 14.1.8 and 14.1.9, does a force that is exerted by one body on another body always need to be in direct contact?**

14.1.11. **Based on your answers to questions 14.1.1 to 14.1.10, explain in your own words the definition of a force?**

Since there are a variety of different types of physical interactions between objects, there are a variety of forces. In general, forces can be placed in two

a)   b)   c)

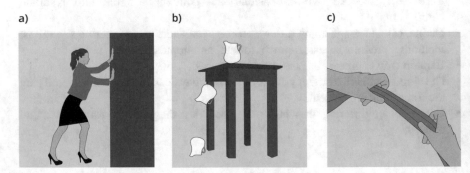

Fig. 14.1 Different types of forces. (**a**) Pushing against a wall (**b**) Jug falling off a table (**c**) Making Taffy

categories; Contact or Applied Forces and Noncontacting forces or Field Forces [1]. Applied Forces are those that result when objects are physically touching each other. Field Forces are those that result from the action at a distance, even though objects are not touching each other. Table 14.1 lists and describes various types of forces.

14.1.12. **For each force listed in Table 14.1, write whether it is considered contact or noncontact force.**

14.2 Tendencies of Forces

Forces can have different effects on objects. When forces are balanced, the net effect is that nothing changes. It is when forces are unbalanced that something happens to the object. The amount of change that occurs depends on the material properties, geometric properties, and support conditions of the object. Understanding the tendencies forces helps engineers to analyze situations and predict behavior. The following set of questions will explore the different tendencies of unbalanced forces.

14.2.1. **If you are pulling a sled across a sheet of ice, like in Fig. 14.2a, what happens to it?**

14.2.2. **If a pull force is placed on an unconstrained object, what happens to it?**

14.2.3. **If you are pulling a slingshot, like in Fig. 14.2b, what happens to it?**

14.2.4. **If a pull/tensile force is placed on a constrained object, what happens to it?**

14.2.5. **If you are stomping on an empty soda can on the ground, like in Fig. 14.2c, what happens to it?**

14.2.6. **If a push/compressive force is placed on a constrained object, what happens to it**

14.2.7. **When playing with and turning a rubrics cube, like in Fig. 14.2d, what happens to it?**

14.2.8. **If an object is constrained and turned, what happens to it**

14.2.9. **When hurricane winds hit a palm tree, like in Fig. 14.2e, what happens to it?**

14.2.10. **What if a force is applied on an object a certain distance away from a point where that object is constrained, what happens?**

14.2.11. **If you pull or push on a playground roundabout (merry-go-round), like in Fig. 14.2f, what happens?**

14.2.12. **If an object is constrained at a center point and a force acts tangent to the fixed center point, what happens to the object?**

14.2.13. **Based on the previous questions, list all of the results that can happen to an object when an unbalanced force act upon it.**

When analyzing an engineering problem, it is good practice to organize the forces into groups, external forces, internal forces, or constraint forces. External

Table 14.1 Different types of forces

Type of force	Description	Associated equation	Examples	Contact or noncontact?
Applied	Touching an object or person can cause many reactions (e.g., push, pull, twist, etc.)	$\vec{F} = m\vec{a}$	Pushing a desk across a room	
Spring	Force exerted by a compressed or stretched spring upon any object that is attached to it	$\vec{F} = k\vec{x}$	Spring attached to screen door to close it	
Gravitational	Small bodies on earth experience a force of gravity that is pulling them toward the center of the earth. On earth, the force of gravity times the mass of an object equals its weight	$\vec{F} = mg$ $g = 9.81 \ \text{m/s}^2$ $g = 32.2 \ \text{ft/s}^2$	Baseball being thrown	
	Larger celestial bodies such as planets and very small objects such as atoms attract each other because of their masses and the distance between them	$\vec{F} = G\dfrac{m_1 m_2}{r^2}$	Satellite orbiting earth	
Normal	Support force exerted upon an object that is in contact with another stable object	Found from FBD	Chair pushing back on a person who is sitting in it	
Friction	Dry force exerted by a surface as an object moves across it or makes an effort to move across it due to irregularities between surfaces	$\vec{F} = \mu N$	Sandpaper	
	Viscous: also known as air resistance, is the force exerted by fluid such as air or water as an object moves through it or makes an effort to move through it	$\vec{F} = \mu A \dfrac{u}{y}$	Water flowing over rocks in a stream	
Electrical	Attractive or repulsive interaction between any two charged objects because of their difference in electrical charge	$\vec{F} = k\dfrac{q_1 q_2}{r^2}$	Rub a rubber balloon against a wool sweater and then touch a person's hair. The result is the hair stands up because it is attracted to the balloon	
Magnetic	Attraction or repulsion that arises between electrically charged particles because of their electron motion	$F = m_e v^2 \dfrac{Q_1 r_e}{r} \dfrac{Q_2}{r}$	Magnets on a refrigerator	

Fig. 14.2 Different tendencies of forces. (**a**) Sled on ice, (**b**) Slingshot, (**c**) Soda can, (**d**) Rubrics Cube (**e**) palm tree in the wind, (**f**) playground roundabout

forces occur between an object or a system and its surroundings. Internal forces act inside a solid structure. Constraint forces act to restrict an object's motion, so it can only move in a certain way.

14.2.14. **For each example below, state whether the force is an external, internal, or constraint.**

- **Wind blowing through the trees**
- **Ball moving along a curved track**
- **Gravity pulling a football to the ground**
- **Tension force holding a bridge cable together**
- **A boat floating on the water due to Buoyancy force**
- **Forces acting at a joint of a roof truss**
- **Box sliding down an incline**

14.3 Force Units and Measurements

The unit of force in the SI system is Newton. It is abbreviated by the symbol capital N. It is a derived unit or secondary unit. Newton is defined based on Newton's Second Law of Motion. In comparison, the USh unit for force is called a pound force. It is abbreviated as lb_f. It is the force applied on a mass of 1 pound (1 lb_m) by the acceleration due to gravity. Recall that in the US unit system, the base unit for mass was called a "slug." Since the SI unit system uses an absolute system of measurement, namely its base values do not depend on gravity, they are acknowledged as the units to be used in scientific measurement. Whereas the US unit system is nonspecific because the force unit can change depending on local gravity; they are rarely used in scientific measurements. This next set of questions helps us to define the Newton unit of force and convert between SI and English force units.

14.3.1. **What is Newton's second law?**

14.3.2. **What are the primary dimensions for Newton's second law?**

14.3.3. **What are the primary SI units for Newton's second law?**

14.3.4. **What is the SI unit definition of a Newton?**

14.3.5. **If a force accelerates a ball with mass of 1 kg at a rate of 1 m/s², how many Newton's of force is felt on the ball?**

14.3.6. **If a force accelerates a ball with mass of 3 kg at a rate of 5 m/s², how many Newton's of force is felt on the ball?**

14.3.7. **If a lb_f is 1 lb_m times the acceleration due to gravity, and if 1 lb_m equals 0.45359 kg, how many Newtons are in 1 lb_f?**

14.3.8. **Based on the previous question, how many lb_f are in 1 N?**

   In almost all engineering disciplines, often forces need to be measured. Based on the definition of a force, an object's mass and acceleration need to be known to measure the force acting on it. However, since we previously learned that forces can have other different effects on objects that information can be used to measure force. Think back to Sect. 14.2, another major effect forces have on objects is that they make objects change shape or deform. A force can stretch, compress, bend, or twist an object. Then if the deformation can be measured experimentally, the force can be deduced.

   A simple example of force measurement is a spring scale. As shown in Fig. 14.3, spring scales are used in everyday life for measuring things like fish size and luggage weight. The device uses the principle of Hooke's Law (named after the scientist who created the law, Robert Hooke). Hooke's law states that for an elastic spring, the force applied to the spring, which causes it to stretch, is directly proportional to the change in length (deformation) of the spring. It can be expressed as the equation $F = kx$; where F is the applied Force, x is the deformation of the spring, and the term k is the spring constant. The spring constant is a term that describes the stiffness of the spring. A stiff spring will have a high spring constant. The spring constant depends on the material that the spring is made from, the shape of the spring, and how the spring is wound. Before beginning measurements, the first thing that needs to be done is to calibrate the spring by measuring the spring stiffness. In order to calibrate a spring scale, one end is fixed to a stand. At the other end of the spring, a known force is applied and the displacement is measured, as shown in Fig. 14.4. This next set of questions helps us learn how to calibrate a spring scale.

14.3.9. **What are the SI and US units of Force?**

14.3.10. **What are the SI and US units of length?**

14.3.11. **Using Hooke's Law and the concept of dimensional analysis and homogeneity (think back to Sect. 10.3), what are the SI units and US units for the spring constant k?**

14.3.12. **For a luggage scale, to determine the value of the spring constant, dead weights are attached to one end of the spring, while the other end**

(a)   (b)

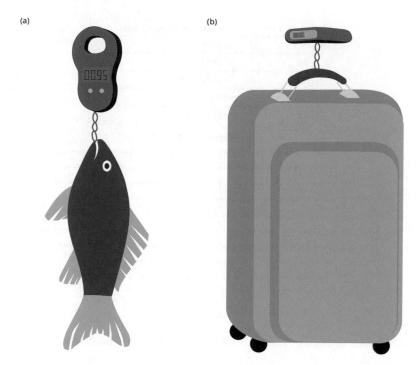

Fig. 14.3 Examples of spring scales being used for (**a**) fish and (**b**) luggage

Fig. 14.4 Calibration of a
Spring Scale. (Picture of
dead weights on spring,
each weight extends it
more)

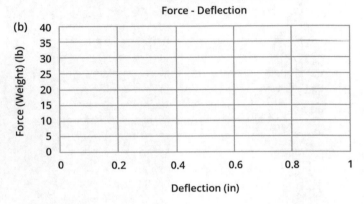

Fig. 14.5 (a) Results of the experiment to calibrate the luggage scale and (b) Graph for results

is fixed to a stand. The measured deflections and weights are shown in the table in Fig. 14.5a. On the blank graph given in Fig. 14.5b, plot out the points from the table. When the values from the table are plotted on the graph, what shape does it make?

14.3.13. What is the equation of the line?

14.3.14. Does the equation of the line look similar to another equation we were just talking about?

14.3.15. Based on the previous question, if plotting Hooke's law on a graph, what type of line would it make?

14.3.16. Based on the previous question, what is the value of k in Hooke's law when plotted on a graph? (Based on all the previous questions, write out a procedure for how to use the data in the table to determine the spring constant) (Hint: use the graph you filled out in Fig. 14.5b.)

14.3.17. Using the data in the table and graph in Fig. 14.5, determine the spring constant for the luggage spring scale.

14.3.18. Rosalind is flying to France for a semester of study abroad, and she wants to check to see if her suitcase is okay to fly (the standard is a suitcase needs to be under 50 lb). She used a luggage scale which has a spring constant the same as found in 14.3.17 and the deflection was 1.3 in. Is her suitcase legal to fly?

While spring scales are a simple, fundamental example, engineers use more precise instruments. Some examples of these devices include force transducers or load cells. An extremely sensitive load cell technology is the atomic force microscope (AFM). This load cell device can measure the forces between groups of atoms. With all the devices the basic principle is the same, namely they measure the deformation of a part of the load cell that acts like a stiff spring, but precision varies depending on the application.

14.4 Force Properties

An important property of Force is that it is a *vector* quantity. To be fully described, it needs both magnitude and direction to be known. The magnitude is typically described by a numerical value representing the amount of force. The direction is usually represented by angles or dimensions in a coordinate system. Whereas a *scalar* quantity only needs magnitude to be known. Typically, engineers' express vectors in two formats, graphically (picture) and analytically (mathematically) [1, 2]

The procedure for representing a vector graphically is as follows:

1. **Choose a coordinate system**. The standard system is typically 2D or 3D Cartesian Coordinates (think back to Chap. 13 on Length). This established our directions with our basis vectors or unit vectors. For Cartesian coordinates, the perpendicular directions are typically denoted as *x-y-z* for 3D and *x-y* for 2D, with the corresponding unit vectors being *i-j-k* in 3D or *i-j* in 2D as shown in Fig. 14.6.
2. **Identify the origin and terminal points**. The beginning point of a vector is typically denoted by a small circle, and many times the vector will begin at the origin of the coordinate system, but that is not always the case. An arrowhead typically denotes the end of a vector.
3. **Display magnitude and direction**. The length of the vector should correspond to the vector's magnitude and should be based on a scale consistent with the overall system. The direction of the arrowhead denotes the vector's *line of action*, which geometrically represents how the force is applied.

The procedure for representing a vector analytically is as follows:

1. **Choose a coordinate system**. This step is the same as above for graphical representation.
2. **Assign coordinate points to the origin and terminal points**. For each end of the vector, a coordinate position can be given. For example, in Fig. 14.6a the beginning of the vector is at the origin of the coordinate system. The same thing can be done with the end of the vector, except now there is a magnitude (F_x, F_y, F_z) in each of the directions.
3. **Report the vector with appropriate notation**. Typical notation for representing a vector in print is to place a straight line bar (–) or arrow (→) over the total vector symbol (such as force $\bar{F}$ or $\vec{F}$), then write it as the sum of the directional components (*x-y-z*) of the force using the magnitudes and unit vectors.

Fig. 14.6 Graphic
representation of a vector
in (**a**) 3D and (**b**) 2D

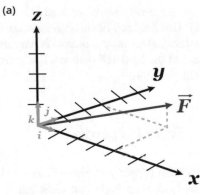

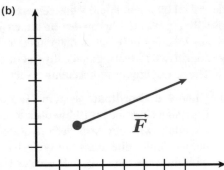

This next set of questions helps practice writing vectors correctly.

14.4.1 **List some examples of scalar quantities.**

14.4.2 **List some examples of vector quantities.**

14.4.3 **Looking at Fig. 14.6a, what is the assigned coordinate point for the beginning of the vector?**

14.4.4 **Looking at Fig. 14.6a, what is the assigned coordinate point for the end of the vector?**

14.4.5 **Based on the previous question, what is your vector magnitude?**

14.4.6 **Based on the previous questions report the full vector shown in Fig. 14.6a with appropriate notation.**

14.4.7 **Looking at Fig. 14.6b, label the directional axes and the unit vectors.**

14.4.8 **Looking at Fig. 14.6b, report the full vector with appropriate notation.**

There are different types of forces depending on where they act with respect to each other, namely their directions.

- *Collinear Forces* all act along the same line of action.
- *Concurrent Forces* are such that their lines of action all pass through the same point in space.
- *Coplanar Forces* are forces that lie in the same plane.

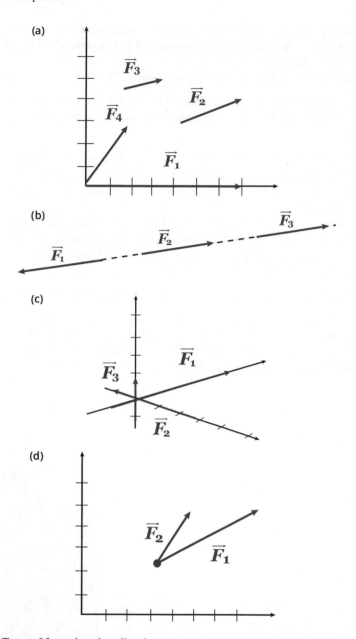

Fig. 14.7 Types of forces based on directions

This next set of questions helps us identify the different types of forces.

14.4.9 **In Fig. 14.7, which forces are all along the same line of action?**

14.4.10 **In Fig. 14.7, which forces have their lines of action all pass through the same point in space?**

14.4.11 In Fig. 14.7, which forces all lie in the same plane?
14.4.12 In Fig. 14.7a through c, Label each of the figures of forces with their appropriate type.
14.4.13 In Fig. 14.7d, how would you label those forces?

Having collinear forces allows engineers to simplify the mathematics associated with the vectors. Specifically, two properties of collinear forces get utilized frequently:

1. The magnitudes of forces acting along the same line can be replaced by a *resultant vector* using algebra. Namely, collinear forces can simply be added together (if they are pointing in the same directions) or subtracted (if they are pointing in opposing directions).
2. When analyzing *external* forces that act on an un-deformable body, the exact point of application does not matter. This is known as the *principle of transmissibility*. It states that a force can be moved along its line of action without altering the external effect that force has on a rigid object.

This next inquiry helps us to appreciate these two properties of collinear forces.

14.4.15 Looking at Fig. 14.8, which forces are acting outside the box?
14.4.16 Looking at Fig. 14.8, which forces are acting inside the box?
14.4.17 Looking at Fig. 14.8, label the internal and external forces?
14.4.18 For the internal forces in Fig. 14.8, are they collinear?

Fig. 14.8 Collinear forces acting on a rigid box

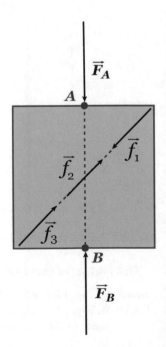

14.4.19 **For the internal forces in Fig. 14.8, write an equation that would replace them with one resultant force. (Assume (+) direction is to the right and up)**

14.4.20 **For the internal forces in Fig. 14.8, if the force values were $\overline{f_1} = 8\,\text{N}$, $\overline{f_2} = 5\,\text{N}$, $\overline{f_3} = 10\,\text{N}$, what is the resultant force?**

14.4.21 **For the external forces in Fig. 14.8, are they the same magnitude?**

14.4.22 **For the external forces in Fig. 14.8, are they in the same direction?**

14.4.23 **For the external forces in Fig. 14.8, with the forces in their current positions, what are they doing to the box?**

14.4.24 **For the external forces in Fig. 14.8, with the forces in their current positions, what is the external result on the box? Why?(Does it get stretched, does it get crushed, does nothing happen?)**

14.4.25 **In Fig. 14.8, draw a picture of what the box would look like if the external forces switched positions.**

14.4.26 **For the external forces in Fig. 14.8, with the forces in the switched positions, what are they doing to the box?**

14.4.27 **For the external forces in Fig. 14.8, with the forces in the switched positions, what is the external result on the box? Why? (Does it get stretched, does it get crushed, does nothing happen?)**

Finally, when a force is fully known (direction, magnitude, line of action) sometimes it is convenient to break up the given force into a number of other vectors. This does not change the net effect of the force. This process is called the **Resolution** of the force. The original force is called the *resultant force*, and the parts it can get broken into are called *components*. The components define the influence of a single vector in that given direction. Opposite to resolution is **Reduction** of forces. In this process, you can take several forces and reduce them into a single resultant force. For both resolution and reduction of forces, trigonometry (recall back to Chap. 11) is used to either deconstruct a force vector into its components or construct a single vector from multiple components. Typically, there are two cases engineers will encounter:

- **Forces in rectangular components**: If two forces are perpendicular to one another, such as in a 2D *xy*-coordinate system, then trigonometry definitions (sine, cosine, tangent) can be used.
- **Forces in non-rectangular components**: When forces are not perpendicular to one another, they are known as non-rectangular components or inclined components. Then resolution or reduction occurs by either triangular laws (Law of Sines or Law of Cosines) or by using the Parallelogram Law of Forces.

This next set of questions helps us to practice resolving and reducing forces.

14.4.28 **In Fig. 14.9a, what coordinate system is the vector in?**

14.4.29 **In Fig. 14.9a, which vector is considered the resultant force?**

14.4.30 **In Fig. 14.9a, which vectors are considered the component forces?**

14.4.31 **In Fig. 14.9a, if you knew the magnitude of the resultant force and the direction (angle θ) that it was pointing at, develop a mathematical**

Fig. 14.9 Resolution and
Reduction of Forces (**a**) in
rectangular coordinates
and (**b**) non-rectangular
components

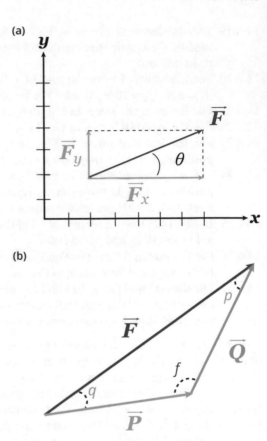

relationship to solve for the magnitude of component forces (hint:
think back to Chap. 11 and trigonometry definitions).

14.4.32 **In Fig. 14.9a, what are the directions of the component forces?**

14.4.33 **In Fig. 14.9a, if you knew the magnitudes of the component forces,
develop a mathematical relationship to solve for the magnitude resultant force (hint: think back to Chap. 11 and trigonometry rules).**

14.4.34 **In Fig. 14.9a, if you knew the magnitudes of the component forces,
develop a mathematical relationship to solve for the direction of the
resultant force (hint: think back to Chap. 11 and trigonometry
definitions).**

14.4.35 **In Fig. 14.9a, if the force $F = 50\,\text{N}$ and $\theta = 30°$, determine its components.**

14.4.36 **In Fig. 14.9b, the vector F is created by adding the two vectors P and
Q. Explain where the head and tail of vector F is placed with respect to
vectors P and Q.**

14.4.37 **In Fig. 14.9b, what shape do the three vectors make?**

14.4.38 **In Fig. 14.9b, if the magnitude of vector Q and the angles f and q were
known, how would you determine the magnitude to resultant vector
F? (Hint: think back to Chap. 11 and trigonometry laws)**

14.4.39 **In Fig. 14.9b, if the magnitude of vector $Q = 8$ in and the angles $f = 120°$ and $q = 30°$, determine the magnitude to resultant vector F. (Hint: think back to Chap. 11 and trigonometry laws)**

14.4.40 **In Fig. 14.9b, if the magnitude of vector F and P are known and the angle f is known, how would you determine the angle p? (Hint: think back to Chap. 11 and trigonometry laws)**

14.4.41 **In Fig. 14.9b, if the magnitude of vector $F = 10$ and $P = 3$ and the angle $f = 120°$, determine the angle p. (Hint: think back to Chap. 11 and trigonometry laws)**

14.4.42 **In Fig. 14.9b, if the magnitude of vector F and P and Q, how would you determine the angle f? (Hint: think back to Chap. 11 and trigonometry laws)**

14.4.43 **In Fig. 14.9b, if the magnitude of vector $F = 8$ in. and $P = 4$ in. and $Q = 5$ in., determine the angle f. (Hint: think back to Chap. 11 and trigonometry laws)**

14.5   Force Parameters

14.5.1   Pressure

When we were discussing length parameters (Chap. 11), we mentioned the length parameter Area, and how important it is in engineering applications. One of the places Area plays a very important role is in the calculation of both pressure and stress on a system. Pressure is measuring the intensity that a force (or set of forces) applies over some designated area. Pressure measurement and calculations can be made for both solid and fluid systems. Please note that pressure is not a force! Rather pressure is the measure that something (fluid or solid) exerts on a surface. The symbol for pressure is commonly P or p. Also, pressure is a scalar value, not a vector quantity [3]. Let us begin by looking at pressure from solid objects. For a solid object, pressure is defined as the force applied perpendicular to the surface of an object divided by the unit area over which that force is distributed. This next set of questions helps us better understand what the magnitude of pressure represents.

14.5.1 **Does it hurt more to get poked in the arm shoulder by a finger or a pin?**

14.5.2 **Figure 14.10 shows a square block of wood ($m = 2.5$ kg) laid on the ground in two different positions. What is the weight of the wood?**

14.5.3 **What is the force that is causing pressure on the ground?**

14.5.4 **Does the wood's weight change depending on the way it is placed on the ground?**

14.5.5 **Does the force change depending on the way it is placed on the ground?**

14.5.6 **Using the definition of pressure given above, for Fig. 14.10a calculate the area that the force is acting upon (hint: if the force is acting down toward the ground, what surface area is perpendicular to that?)**

(a)   (b)

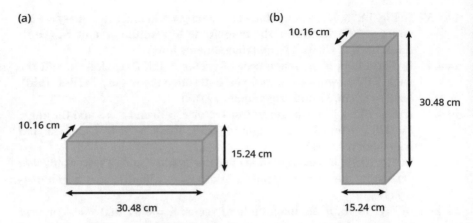

Fig. 14.10 A solid block of wood with dimensions given (**a**) resting on its side and then and (**b**) resting on its end

14.5.7 Using the definition of pressure given above, for Fig. 14.10a calculate the pressure at the contact surface.

14.5.8 Using the definition of pressure given above, for Fig. 14.10b calculate the area that the force is acting upon (hint: if the force is acting down toward the ground, what surface area is perpendicular to that?)

14.5.9 Using the definition of pressure given above, for Fig. 14.10b calculate the pressure at the contact surface.

14.5.10 Does the pressure change depending on the way the wood is placed on the ground?

14.5.11 Which picture in Fig. 14.10 has the largest area?

14.5.12 Which picture in Fig. 14.10 has the smallest area?

14.5.13 Which picture in Fig. 14.10 has the highest pressure?

14.5.14 Which picture in Fig. 14.10 has the smallest pressure?

14.5.15 Based on the previous set of questions can you come up with a general guideline for the relationship between contact area and pressure?

14.5.16 What are the standard SI and US units of pressure?

The previous set of questions investigated pressure derived from a solid object. Pressure can also develop from interactions with fluids (vapor or liquid). There are two common scenarios for fluid pressures to form:

1. Static or "non-moving" conditions. This typically refers to large open bodies of fluid such as a lake, swimming pool, the open, or the atmosphere. In these scenarios, even though the fluid is moving (with waves, wind, etc.) the body of fluid is so large that those types of changes in pressure can be neglected.

2. Dynamic or "moving" conditions. This typically refers to fluid moving through a container of some sort (a pipe, nozzle, etc.)

Let us briefly investigate static fluid pressure. There are two straightforward but important laws that govern fluids are rest. The first states that the pressure of a fluid acting at any point within the fluid is equal in all directions (this is known as Pascal's principle) [4]. The second states that the pressure of a fluid increases as the distance into the fluid increases. The relationship between the pressure and the height of a column of fluid is given by $P = \rho g h$, where ρ is the density of the fluid (units of kg/m^3 or slugs/ft^3), g is the gravitational constant, and h is the height of the column of fluid. This next set of questions lets us explore the pressure in static fluids.

14.5.17 **When you swim at the bottom of the deep end of a swimming pool, what happens to your head and ears?**

14.5.18 **You are in a swimming pool and you swim to the bottom of the shallow end which is 4 ft deep. Calculate the pressure your body feels (density of water is 1.94 slugs/ft^3).**

14.5.19 **You are in a swimming pool and you swim to the bottom of the shallow end which is 12 ft deep. Calculate the pressure your body feels (density of water is 1.94 slugs/ft^3).**

14.5.20 **In a swimming pool, where is the pressure the highest (the shallow end, the deep end, or the pressure is the same everywhere)?**

14.5.21 **What is the depth (in feet) you would need to swim to in the ocean to feel 20 psi of pressure? (density of water is 1.94 slugs/ft^3)**

When looking at an enclosed fluid system we can again use Pascal's principle. Since the pressure of the fluid at the bottom of an enclosed container will be equal to the pressure at the top of the container, when you apply a force to the fluid, it will be equally distributed and undiminished. Therefore, if the geometry of the enclosed container is not constant a phenomenon known as multiplication of forces occurs. This concept is applied to hydraulic systems (such as Hydraulic Lifts, Jacks, and Breaks) that are used for holding, lifting, or moving large loads with a small amount of applied force. This next set of questions helps us formulate and apply a general relationship between pressure, force, and area in a hydraulic system.

14.5.22 **Write a mathematical equation that relates pressure to force and surface area.**

14.5.23 **Looking at Fig. 14.11, write an equation for the pressure felt at the first piston in terms of the F_1 force and the surface area A_1.**

14.5.24 **Looking at Fig. 14.11, write an equation for the pressure felt at the second piston in terms of the F_2 force and the surface area A_2.**

14.5.25 **Looking at Fig. 14.11, due to Pascal's law, when you apply force F_1, when the pressure transfers through the fluid to the second piston, does the pressure increase, decrease, or remain the same?**

14.5.26 **Write a mathematical expression that displays the answer of the previous question.**

14.5.27 **Using questions 14.5.22 through 14.5.26, develop an equation for the force felt by the larger piston based on an input force at the first piston.**

14.5.28 **For the system in Fig. 14.11, if the pistons were cylindrical shaped ($R_1 = 5$ cm, $R_2 = 20$ cm) and a force was placed on piston 1 of 5 N, what would be the force felt at piston 2?**

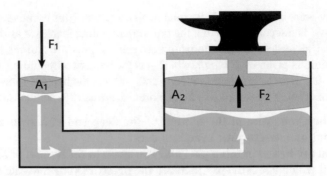

Fig. 14.11 Example of simple hydraulic lift. There are two pistons on either side of the container and the container is filled with water (which is incompressible). According to Pascal's law, the pressure applied to the small piston due to the first force will be transferred equally and undiminished by the large piston

14.5.29 **Based on the previous question does the force felt at the second piston increase, decrease, or remain the same?**

14.5.30 **Using the system shown in Fig. 14.11 where the pistons are cylindrical shaped ($R_1 = 1$ m, $R_2 = 6$ m), if a 100 kg mass was placed on piston 1, develop a process to figure out if the anvil on piston 2 could be lifted. (Note: the anvil's mass is roughly 3400 kg. Hint: Think about what is causing the force on the pistons?)**

14.5.31 **Using the system shown in Fig. 14.11 where the pistons are cylindrical shaped ($R_1 = 1$ m, $R_2 = 6$ m), if a 100 kg mass was placed on piston 1, could the anvil on piston 2 be lifted up? (Note: the anvil's mass is roughly 3400 kg. Hint: Think about what is causing the force on the pistons?)**

   Finally, when dealing with pressure due to vapors or gases many standard measurements are not absolute. This means that many gas pressure measurements are made relative to a specific reference. The references are referred to as zero references and they are typically stated in parenthesis following the unit. The following are several important references for engineers to know:

• **Atmospheric Pressure** (also known as barometric pressure)—This is the pressure due to the weight of the air within the Earth's atmosphere. Think of it as a column of air pushing down on your head that extends all the way to the Exosphere. The unit for atmospheric pressure is called the *standard atmosphere* (symbol: atm), and at sea level it is approximately 1 atm.

• **Gauge Pressure**—This is a measurement of pressure in a system relative to atmospheric pressure. Since it accounts for atmospheric pressure, it is positive for pressures above atmospheric pressure and negative for pressures that are below atmospheric pressure.

• **Absolute Pressure**—This measurement is set with reference to a perfect vacuum. Namely, if there were no air pressure at all. Then since on earth there is almost always atmospheric pressure, Absolute Pressure is equal to the gauge plus atmospheric pressure.

- **Vapor Pressure** (also known as equilibrium vapor pressure)—This is an indicator of how fast a fluid will evaporate to gas under certain atmospheric conditions (such as temperature). One way to think about it is that vapor pressure is the pressure (under ambient temperature) that the substance needs to be kept at in order for it to remain in liquid state. The higher the vapor pressure of a substance at normal temperature, the faster it evaporates. For substances with low vapor pressure under normal conditions, the slower is evaporation.
- **Ideal Gas Pressure**—If a vapor is considered an ideal gas, then the molecules in that gas have essentially no volume and intermolecular forces interacting with each other. This allows engineers to use a simplified equation of state to model many situations. The law states that the pressure (P) of an ideal gas is directly proportional to the amount of substance the gas has (n) and the temperature (T) and is inversely proportional to the gas' volume (V). In mathematical form, it is $PV = nRT$, where R is the ideal gas constant.

Pressures are typically expressed in units of Pascal (Pa), standard atmosphere (atm), pounds per square inch (psi), millimeters of mercury (mmHg), and inches of mercury (in.Hg). This next inquiry helps us to become familiar with pressure references and units.

14.5.32 **Looking at Table 14.2, what is the altitude at sea level?**

14.5.33 **Looking at Table 14.2 and based on the above definition of Atmospheric Pressure, what is altitude origin (namely, where do we start measuring the altitude from) for one standard atmosphere?**

14.5.34 **Looking at Table 14.2, what is the atmospheric pressure at sea level?**

14.5.35 **Looking at Table 14.2 and based on the previous questions, what is a standard atmosphere (1 atm) in kilopascals?**

14.5.36 **Looking at Table 14.2, compared to sea level, does the atmospheric pressure at the top of the Burj Khalifa increase, decrease, or remain the same?**

14.5.37 **Looking at Table 14.2, compared to sea level, does the air density at the top of the Burj Khalifa increase, decrease, or remain the same?**

Table 14.2 Variation of atmospheric pressure and density with altitude

Location	Altitude (m)	Altitude (ft)	Atmospheric pressure (kPa)	Air density (kg/m³)
Sea level	0	0	101.325	1.225
Burj Khalifa (tallest building on earth)	828	2717	91.79	1.1309
Denver, Colorado	1500	5000	84.55	1.058
Mount Everest	8848	29,029	30.80	0.467
Commercial Airplanes average cruising altitude	11,000	36,089	22.70	0.365

14.5.38 Looking at Table 14.2, what is the altitude atmospheric pressure and density on the top of Mount Everest?

14.5.39 Looking at Table 14.2, what percentage of atmospheric pressure does the top of Mount Everest have compared to sea level?

14.5.40 Looking at Table 14.2, what percentage of air density does the top of Mount Everest have compared to sea level?

14.5.41 Based on the previous questions, what trend do you see with Atmospheric Pressure? Explain why this happens?

14.5.42 Based on the previous questions, what trend do you see with air density? Explain why this happens?

14.5.43 Looking at Table 14.2, explain why airplane cabins are pressurized during flights.

14.5.44 Based on the explanation of atmospheric, absolute, and gauge pressure, write a mathematic expression for absolution pressure in terms of atmospheric and gauge pressure.

14.5.45 When reading tire pressure (in bike or car tires) the measurement is the gauge pressure. If a tire gauge pressure measurement reads 32 psi (lb/in.^{2}), explain what that reading means in relation to the atmospheric pressure.

14.5.46 If a tire gauge pressure measurement at sea level read 32 psi, what is the absolute pressure inside the tire (note: 1 psi = 6895 Pa)?

14.5.47 If a tire gauge pressure measurement in Denver, Colorado read 32 psi, what is the absolute pressure of the air inside the tire?

14.5.48 What does it physically mean if the gauge pressure reading is negative?

14.5.49 What is the relationship between the pressure and the height of a column of fluid (Hint: look back to the second paragraph of this section)?

14.5.50 What is the atmospheric pressure at sea level in terms of Newton per meter square (Recall this is the pressure created at the bottom of a column of air at sea level)?

14.5.51 What is the gravitational constant in SI units?

14.5.52 Based on the previous three questions, how high would a fluid column of mercury need to be in order to have the same standard atmospheric pressure as air at sea level? (density of mercury is 13,550 kg/m^3) (Put your final answer in millimeters)

14.5.53 Based on the previous question, write the unit conversation relationship between standard atmospheres and mmHg.

14.5.54 If 1 mm = 0.03937 in., what is the unit conversation relationship between standard atmospheres and in.Hg?

Understanding stationary and moving fluid pressures is important in many aspects of engineering. For example, hydrostatics is very important in the engineering of large and small fluid-containing structures (such as damns and oil barrels). As engineers, you will investigate these concepts and many more in the fundamental classes such as fluid mechanics, dynamics, and aerodynamics.

14.5.2 *Stress*

In Sect. 4.2, when investigating the tendencies of forces, we discovered that forces can both move and *deform* objects. When designing a machine part or structure, it is important for engineers to calculate all the forces acting on an object and determine if the object is going to change shape due to those forces. For example, if a rod is being pulled on by an external mass like in Fig. 14.12a, internally the force is trying to overcome the bonds that are holding the rod together and separate the individual atoms of material. Engineers quantify this by a term called stress. **Stress** provides a measure of the strength of a force as it acts over an area. There are typically two main types of stress, normal stress, which is designated with the Greek letter σ and shear stress which is designated with the symbol τ. Normal stress (σ) is defined as force acting perpendicular to the are,a and shear stress (τ) is defined as force acting parallel to the area This next set of questions helps us define stress and identify the types of stress typically seen in solid objects.

14.5.55 **Describe what is happening to the rod in Fig. 14.12a.**

14.5.56 **Which way is the force acting in Fig. 14.12a, parallel to the bar surface or perpendicular or the bar surface.**

14.5.57 **What is the shape of the area the force is acting over?**

14.5.58 **Based on Fig. 14.12a how would you define stress (force acting over area).**

14.5.59 **Based on the definitions in the above paragraph, would Fig. 14.12a be normal or shear stress?**

14.5.60 **Describe what is happening to the rod in Fig. 14.12b.**

14.5.61 **Which way is the force acting in Fig. 14.12b, parallel to the bar surface or perpendicular or the bar surface?**

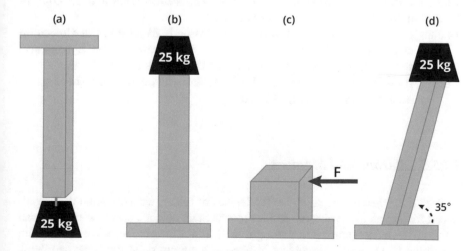

Fig. 14.12 A rod subject to a mass (or force) in different positions

14.5.62 What is the shape of the area the force is acting over?

14.5.63 Based on Fig. 14.12b how would you define stress?

14.5.64 Based on the definitions above, would Fig. 14.12b be a normal or shear stress?

14.5.65 Describe the two effects that normal stress can have on an object.

14.5.66 Label the stress states in Fig. 14.12a, b as either normal tensile stress, normal compressive stress, or shear stress.

14.5.67 If I grab a grilled cheese sandwich by the outside bread and slide the bread in opposite directions, what will happen to the filling?

14.5.68 In Fig. 14.12c, which way is the force acting, parallel to the surface or perpendicular or the surface?

14.5.69 In Fig. 14.12c, what is the shape of the area the force is acting over?

14.5.70 Based on Fig. 14.12c, how would you define stress?

14.5.71 Based on the definitions above, would Fig. 14.12c be normal or shear stress?

14.5.72 Label the stress states in Fig. 14.12c as either normal tensile stress, normal compressive stress, or shear stress.

14.5.73 Based on the previous definition of stress, what are the units of stress in SI and US systems?

14.5.74 In Fig. 14.12a if the mass in the figure is 25 kg and the dimensions of the bar are 15 mm by 15 mm, calculate the stress in the rod.

14.5.75 Looking at Fig. 14.12d, is the force acting vertically, horizontally, or both?

14.5.76 Is force a vector?

14.5.77 Based on the previous question and the definition of stress, is stress a vector?

14.5.78 Based on the previous question and what you learned in Sect. 14.4, what can we do to the force vector in Fig. 14.12d?

14.5.79 Based on the previous question, in Fig. 14.12d what are the components of stress acting on the surface?

14.5.80 Looking at Fig. 14.12d, if the force acting is 25 kg and the dimension of surface it is sitting on is 15 mm by 15 mm, calculate the normal and shear components of the force (hint: think back to Sect. 14.4).

14.5.81 Based on the previous question, calculate the normal and shear stress acting on the rod if Fig. 14.12d.

14.5.3 Torque and Moment

In Sect. 6.2, when investigating the tendencies of forces, we discovered that unbalanced forces can move objects. If a force acts at an object's center of mass, it can translate the object (move it in a straight line). If however the force acts away from the center of mass or if the object is fixed at a certain point the force can bend, rotate, or twist the object. This leads engineers to use two terms to describe the results of a force that acts a distance away from a particular point on an object. The two terms used are *Torque* and *Moment*.

The term Torque is used when the result of the object is twisting. **Torque** measures the force that causes an object to rotate around a specific axis. Similar to linear acceleration being created by a force ($F = ma$), angular acceleration is created by torque. Torque is defined as a force multiplied by a distance. Not just any distance though. The distance needs to be *perpendicular* to the line of action of the force. This is also sometimes known as the moment arm. The symbol for torque is either capital T or the Greek letter tao (τ).

When the force results in the object bending or rotating, it is referred to as a **Moment**. The term Moment typically is used for static, nonrotational situations, such as in the analysis of forces acting on a beam. The symbol for moment is capital M. For both moments and torques, the standard procedure is to label a moment/torque *positive* if the tendency of the force is to rotate the object *counterclockwise*, and label a moment/torque *negative* if the tendency of the force is to rotate the object *clockwise*. This next set of questions lets us explore calculating torques and moments.

14.5.82 **Based on the definition above, write a mathematical formula for torque.**

14.5.83 **What are the fundamental dimensions for torque?**

14.5.84 **What are the SI and US units for torque?**

14.5.85 **Looking at the door from the front shown in Fig. 14.13a,** what is the force F_1 doing to the door handle?

14.5.86 **Looking at Fig. 14.13a, what is perpendicular distance to the center of rotation?**

14.5.87 **Looking at Fig. 14.13a, which way is the force turning the door handle?**

14.5.88 **Looking at Fig. 14.13a, is the force creating a positive or negative torque?**

14.5.89 **Looking at Fig. 14.13a, calculate the torque that is put on the door handle if F_1 is 18 N and the door handle diameter is 6 cm.**

14.5.90 **Looking at the door from the top shown in Fig. 14.13b, when you open the door with force F_2, which way does the door swing, counterclockwise or clockwise?**

14.5.91 **Looking at the door from the top shown in Fig. 14.13b, when you open the door with force F_2, is the force creating a positive or negative moment?**

14.5.92 **Looking at Fig. 14.13b, is the line of action of force F_2 perpendicular to point A?**

14.5.93 **Looking at Fig. 14.13b, does force F_2 cause a moment at point A?**

14.5.94 **Looking at Fig. 14.13b, what is the moment of force F_2 at point A?**

14.5.95 **Looking at Fig. 14.13b, is the line of action of force F_2 perpendicular to point C?**

14.5.96 **Looking at Fig. 14.13b, does force F_2 cause a moment at point C?**

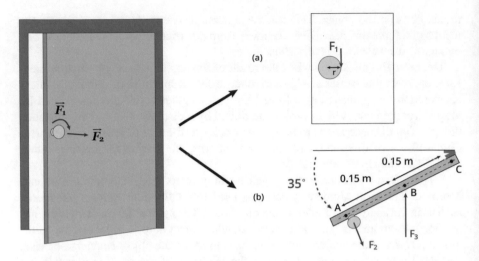

Fig. 14.13 Moments and Torques created by opening a door. (**a**) View from the front and (**b**) view from the top

14.5.97 **Looking at Fig. 14.13b, what is the moment of force F_2 at point C if F_2 is 100 N?**

14.5.98 **Looking at Fig. 14.13b, when you push on the door with force F_3, which way does the door swing, counterclockwise for clockwise?**

14.5.99 **Looking at Fig. 14.13b, can the force F_3 being applied to the door be broken into horizontal and vertical components? If so, what are they?**

14.5.100 **Looking at Fig. 14.13b, can the distance from point B to point C be broken into horizontal and vertical components? If so, what are they?**

14.5.101 **Is force a vector?**

14.5.102 **Based on the previous questions, are torque and moment vectors? Explain why?**

14.5.103 **Looking at Fig. 14.13b, which is the perpendicular distance from the force at point B to point C? The horizontal or vertical component? (Answer: the horizontal component)**

14.5.104 **Looking at Fig. 14.13b, what is the moment of force F_3 at point C if F_3 is 100 N?**

14.5.105 **Based on the definition of a moment, explain why trees have trunks that are thickest near the ground and thin branches that extend off the trunk?**

14.5.4 Mechanical Work

Work is associated with energy transfer (we will talk about energy in more detail in Chap. 17). Mechanical work specifically addresses the energy transfer from a force. **Mechanical work** is done when a force moves an object over some distance.

Mechanical work can be calculated by multiplying a constant force by the distance over which it moves an object. The force acting is parallel to the distance. Even though force is a vector and can be resolved into components, work is a scalar. The negative or positive value of work indicates energy transfer. Positive work is when a force results in transferring energy to an object and negative work is when the force results in removing energy from an object. Mechanical work is represented by the symbol W. These next questions help us to describe and calculate mechanical work.

14.5.106 **Based on the information given in the previous paragraph, write a mathematical equation for work.**

14.5.107 **What are the primary dimensions of work?**

14.5.108 **What are the SI and US units of work?**

14.5.109 **A baseball pitcher throws a baseball to a catcher. Is the pitcher placing a force on the ball to transfer energy into the ball or remove energy from the ball?**

14.5.110 **A baseball pitcher throws a baseball to a catcher. Is the pitcher doing positive or negative work?**

14.5.111 **A baseball pitcher throws a baseball to a catcher. Is the catcher placing a force on the ball to transfer energy into the ball or remove energy from the ball?**

14.5.112 **A baseball pitcher throws a baseball to a catcher. Is the pitcher doing positive or negative work?**

14.5.113 **Looking at Fig. 14.14 of a lawnmower being pushed, can the force F be broken into horizontal and vertical components? If so, what are they?**

14.5.114 **Looking at Fig. 14.14 of a lawnmower being pushed, which component of the force is contributing to the movement of the mower, the horizontal or vertical component? What is the other component doing?**

14.5.115 **Looking at Fig. 14.14 of a lawnmower being pushed, if the mower gets pushed 15 ft and the mower is being pushed with 10 lb of force, what is the work that is being done mowing the lawn?**

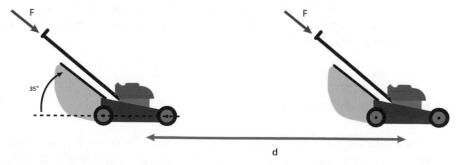

Fig. 14.14 The work done on the lawnmower by force F as it travels over distance d

14.5.116 **You are pushing on the side of a college dorm building's brick wall and it does not move. Is there mechanical work being done? Explain why or why not?**

14.5.5 Linear Impulse

The previous few sections investigated the results of force acting at a distance (moment/torque) or through a distance (work). Another important parameter of force describes how force effects something over a period of time. **Linear Impulse** is defined as the total effect of a force acting over a period of time. This linear impulse results in a change in the linear momentum of an object. In Chap. 13 we learned about the mass parameter linear momentum. It was quantified as the mass multiplied times the velocity and it described mass in motion. So the relationship between linear impulse and momentum can be written as:

$$\text{Linear impulse}\left(\text{force over time}\right) = \text{Momentum}_{\text{final time}} - \text{Momentum}_{\text{initial time}}$$

$$\vec{F}_{ave}\Delta t = m\vec{v}_{\text{final}} - m\vec{v}_{\text{initial}}$$

where F is the average of the magnitudes of the forces acting on the object, Δt is the period of time over which the forces acts on the object, m is the object mass, and v is the initial and final velocities of the object. It is important for engineers to have a good understanding of linear impulse forces when designing safety equipment for impact events such as football and lacrosse helmets and pads, car airbags, or packaging for electronics in case of drops. Engineering students typically study impulse in detail in a vector mechanics dynamics class. These questions help us to understand the linear impulse.

14.5.117 **Based on the definition given, what are primary dimensions of linear impulse?**

14.5.118 **What are the SI and US units of linear impulse?**

14.5.119 **During a softball game, a pitcher throws a pitch at 80 ft/s. The ball weighs 0.5 lb. If a softball player hits a line drive straight back to the pitcher with a force of 95 lb over 0.015 s, how fast does the ball travel back to the pitcher?**

14.6 Free Body Diagrams

One of the most important skills engineers can develop is visualizing forces; seeing where they act and the response that occurs. One way this is accomplished is through a **Free Body Diagram (FBD)**. Engineers draw Free Body Diagrams so that they can visualize where forces are acting on a body, then they can easily set up equilibrium equations to solve for desired unknowns. A FBD is a drawing of a singular object where all of the external forces, supports, connectors, etc., are turned into vector force arrows. The arrows show the magnitude and direction of each force vector with respect to a coordinate system. The skill of drawing Free Body Diagrams will be used in classes such as physics, statics, dynamics, solid mechanics, and design and failure analysis of machine elements. Here are the recommended steps for constructing a Free Body Diagram:

1. **Isolate the body of interest**. This means that you only draw the object that you are looking at. Supports and connectors are not included (they will later be replaced with vector arrows). Generally, it is customary for an object in a FBD to be simplified, for example representing the object as a square.
2. **Draw the axis for the system**. Typically the axis of the system is labeled at the center of mass of the object or the square (as long as you assume a constant mass object). The engineer can choose the coordinate system (such as Cartesian, Polar, or Path coordinates) such that the resolution of the forces is mathematically convenient. Choosing coordinate systems take practice and experience.
3. **Add in the applied forces as vectors**. Draw all the external forces as arrows pointing in the direction that force is acting. Typical forces such as the object's weight, and any pull or push force need to be considered. The arrows are then labeled with a symbol or a numerical value (if one can be calculated).
4. **Replace supports with force vectors**. Replace things that are supporting or restraining the object with forces. Typically the ground or a floor is replaced by a normal force. Or if the object is also horizontally translating a friction force will need to be added. There is no set rule about the number of forces that must be drawn in a free-body diagram. It depends on the system and the assumptions the engineer is making.
5. **Include appropriate geometric or dimensional information**. For the final step, include any data that is going to be needed in the resolution of vector calculations. This includes geometric dimensions (such as object size or radius) that could be used in moment calculations. Then dimensional information such as angles will help resolve any force vectors into their components that do not lie along the chosen coordinate directions.

This set of questions gives us practice identifying forces and drawing Free Body Diagrams.

14.6.1 **Looking at Fig. 14.15, if drawing a FBD of block 1, what shape would you draw?**

14.6.2 **Looking at Fig. 14.15, if drawing a FBD of block 1, what coordinate system would you use, Cartesian or something else? Explain why.**

14.6.3 **Looking at Fig. 14.15, if drawing a FBD of block 1, how would you orient your coordinate axis, vertical and horizontal or along the angle of the incline?**

Fig. 14.15 Depiction of
bodies connected by a
string, a classic mechanics
problem. A free body
diagram can be created for
each mass

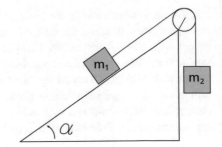

14.6.4 **Looking at Fig. 14.15, if drawing a FBD of block 1, name the applied
forces shown.**

14.6.5 **Looking at Fig. 14.15, if drawing a FBD of block 1, name the support forces.**

14.6.6 **If you oriented your axis horizontal and vertical, which vector forces
are not aligned with the coordinate system? Which vector force(s) are
aligned with the coordinate system?**

14.6.7 **If you oriented your axis along the angle of the incline, are which vector force(s) are not aligned with the coordinate system? Which vector
force(s) are aligned with the coordinate system?**

14.6.8 **Referring to all the previous questions, draw and label a FBD of block
1 (remember to include the angle dimensions of any vectors that are
not aligned with your chosen coordinate axes).**

14.6.9 **Draw a FBD of block 1 if it was assumed that the incline was slippery
so that there was no friction.**

14.6.10 **Draw a FBD of block 2.**

End of Chapter Questions

IBL Problems

IBL1:

1. What is the length of the parallelogram sides in terms of the force vectors $\vec{P}$ and $\vec{Q}$?
2. What is the length of the resultant $\vec{R}$ in terms of the parallelogram?
3. Consider the triangle formed by ACE, is it a right triangle?
4. Looking at triangle ACE, develop a relationship between the lengths of the sides (hint: think back to our discussion on length and trigonometry).
5. Consider the line AE, develop a relationship between the total length of the line and the length of its fragments.
6. Using the previous two questions, remove the total length of line AE from the equation you developed in question 4 and fully expand the answer.
7. Consider the triangle formed by BCE, is it a right triangle?
8. Looking at triangle ABCE, develop a relationship between the lengths of the sides (hint: think back to our discussion on length and trigonometry).
9. Using the previous answer, remove the total length of line EC from the equation developed in question 6 and simplify the answer.
10. For the equation developed in question 9, replace the line lengths with the vector representations (hint: look back at questions 1–2). Is there one length term that cannot be removed (there is no direct vector relationship for it)? What is it?
11. Using triangle BCE, represent line BE in terms of a vector representation (hint: what kind of triangle is BCE and what vector and angle are there?)
12. Using the previous answer, replace the term in question 10 to create the final equation. (Note: this is called the Parallelogram Law of Forces.)
13. If two forces 5 N and 25 N are acting at an angle of 120° between them, Find the magnitude of the resultant force.

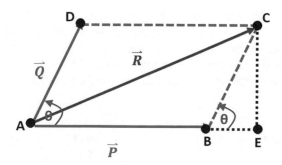

IBL2:

1. If the elevator is going up with a constant acceleration, draw a Free Body Diagram of the person.
2. What is an equation for the force due to weight of the person?
3. What is the force that the scale is reading?
4. Write Newton's second law in terms of the forces that are acting on the person as the elevator is going up.
5. Using the previous three questions, write an equation for the force read by the scale based on the weight of the person and the travel of the elevator.
6. If the elevator was traveling up at 3 m/s², what is the force read in the scale?
7. If the elevator is going down with a constant acceleration, draw a Free Body Diagram of the person.
8. Write Newton's second law in terms of the forces that are acting on the person as the elevator is going down.
9. Using the previous three questions, write an equation for the force read by the scale based on the weight of the person and the travel of the elevator.
10. If the elevator was traveling down at 3 m/s², what is the force read in the scale?
11. Explain why there is a different reading on the scale when the elevator is going up compared to when it is going down.

Practice Problems

1. Determine the sum of the moments of the forces about point A shown in Fig. 14.16.
2. A person has a mass of 62 kg. What is the weight of that person on Mercury ($g = 3.7$ m/s²), the moon ($g = 1.62$ m/s²), and on earth in both SI and English units?
3. The current world record for the extreme sport of free diving is 214 m below the water surface. Calculate the absolute exerted by the water on a free diver at that depth.
4. A hydraulic system like the one in Fig. 14.11 has the radial dimensions $R_1 = 2$ in. and $R_2 = 8$ in. If a load of 4000 lb is placed on the second platform, how much force is needed on the first platform to hold the load up steady?
5. If you have 1000 lb hanging from a rod as in Fig. 14.12a, the cross section of the rod is a square with sides of length 3 in. and length 7 in. Calculate the stress seen in the rod.

Fig. 14.16 Beam and loads for question 1

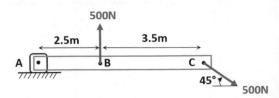

6. Draw a Free Body Diagram for the following objects shown in the figure:

   - The square weight at the end
   - The pallet holding the round boulder
   - The pulley holding the boulder and pallet
   - The person pulling the rope

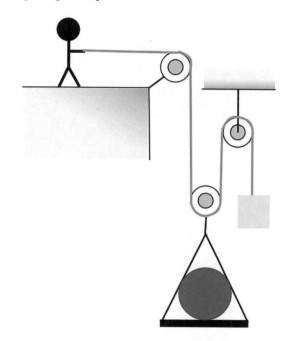

References

1. Bower, Allan F. (2010), Applied Mechanics of Solids, Taylor and Francis ISBN-13: 978-1439802472.
2. Eide A.R., Jenison R.D., Mickelson S.K., Northrup L.L. (2018), Engineering Fundamentals and Problem Solving. McGraw Hill, ISBN: 978-0-07-338591-4.
3. Knight, PhD, Randall D. (2007). "Fluid Mechanics". Physics for Scientists and Engineers: A Strategic Approach (google books) (2nd ed.). San Francisco: Pearson Addison Wesley. ISBN 978-0-321-51671-8.
4. "Pascal's principle - Definition, Example, & Facts". britannica.com. Viewed 21 September 2020.

Chapter 15
Temperature

Abstract Temperature is an important quantity that affects many aspects of our daily lives. It is also very important for engineers to understand because it is one of the seven fundamental dimensions that describe our physical world. In this chapter, first you will explore the importance of temperature in engineering and our everyday lives. Next, you will learn about temperature measurement and temperature scales, paying particular attention to converting between scales. Next, we will explain the difference between temperature and heat transfer and examine methods of heat transfer. Finally, we will discuss a variety of temperature-related properties that effect materials.

By the end of this chapter, students will learn to:

- Define temperature and explain the importance of the notion of temperature in everyday life and in engineering analysis.
- List examples of temperature and related variables.
- Explain the different scales of temperature.
- Convert between different temperature scales.
- Explain the importance of relative temperature difference versus absolute measurement.
- Define heat transfer.
- Explain the distinction between temperature difference and heat transfer.
- Identify the common units of heat transfer for each unit system and be able to convert between them.
- Identify methods of heat transfer and calculate heat transfer for the different modes.

15.1 Importance of Temperature Parameters

Recall in our previous chapters that humans needed ways to describe their physical surroundings (dimensions), and then quantify their observations (units) to better understand their surroundings. Temperature is the fundamental quantity that helps

M. Blum, *An Inquiry-Based Introduction to Engineering*,
https://doi.org/10.1007/978-3-030-91471-4_15

living beings describe how hot or cold something is. But there is more to it than just that. Temperature is so important in describing the various states of things. Have you ever asked yourself questions like "Is it cold enough outside so that I need to wear a jacket?" or "Why did my phone overheat while it was charging?" You have probably asked hundreds of similar temperature-related questions throughout your life. Developing a good understanding of what is meant by temperature and how it is quantified is necessary for ALL engineering disciplines. This inquiry helps us define and explore the importance of temperature and has us identify how it effects our lives and its importance in engineering. The symbol for temperature is typically noted as a capital T.

15.1.1 **Give three more examples (other than the one given above) of questions that involve temperature.**

15.1.2 **When liquid water turns to ice, do the particles increase in speed or decrease in speed?**

15.1.3 **Atoms that are slow moving, do they have high or low temperature?**

15.1.4 **When the temperature of an object decreases, does its temperature increase or decrease?**

15.1.5 **When liquid water turns to vapor, do the particles increase in speed or decrease in speed?**

15.1.6 **Atoms that are fast moving, do they have high or low temperature?**

15.1.7 **When the temperature of an object increases, does its temperature increase or decrease?**

15.1.8 **Based on the previous questions, in your own words, how would you define temperature?**

15.1.9 **Based on the previous question, in your own words explain why the concept of temperature is important.**

15.1.10 **Figure 15.1 shows some ways temperature affects us. Give three examples of situations from everyday life where temperature plays an important role.**

15.1.11 **Give three examples of situations from engineering where temperature plays an important role.**

15.2 Measure of Temperature

Throughout history, as well as many times in our current life, humans relied on the senses of vision and touch to measure temperature. For example, before jumping into a lake, a person might dip their foot in so as to gauge how cold the water is. Also, we tell children never to touch the stovetop when the burners are glowing red. However, these methods have limitations, specifically they are limited in accuracy and they cannot universally quantify a value for a temperature. You cannot stick your foot in a lake and tell that the temperature is exactly 69.7 °F. So, in order to measure temperature humans needed to develop a device that provided information

Fig. 15.1 Everyday examples of how temperature effects our lives

about temperature that was quantifiable, accurate, and could be understood by everyone. This led to the invention of thermometers.

The most common type are bulb thermometers. This device measures temperature based on the thermal expansion or contraction of a fluid (such as alcohol or liquid mercury). When the temperature increases, the liquid expands its volume and rises in a small glass capillary tube. When the temperature decreases, the liquid contracts its volume and falls in the tube. With the progress of society and advancements in technology came the invention of other types of thermometers such as bimetallic strip and infrared. In order to enumerate thermometers, different temperature scales were developed. Each with their own calibration system. There are a few different temperature scales, but these are the four most common that are used by engineers:

- **Fahrenheit** (°F): This scale was developed by Daniel Fahrenheit, a Polish physicist and inventor. He calibrated his measurement system deciding that the freezing and boiling points of water will be 180° apart, in order to eliminate fractions and make the units more easily subdivided [1]. The freezing point of water was marked at 32 °F, and the boiling point was marked at 212 °F. This scale is the primary scale for everyday use in mainly the United States. Most other countries use the Celsius scale.
- **Rankine** (°R): This scale was created by Scottish mechanical engineering and physicist William John Macquorn Rankine [2]. This temperature scale was based on the Fahrenheit temperature scale, except it is calibrated so that its zero degree corresponds to the value of *absolute zero*. *Absolute zero* is the lowest limit of the thermodynamic temperature scale. At this value, which has never been experimentally reached, particles have essentially no kinetic motion.
- **Celsius** (°C): Originally known as the centigrade scale, this scale is the most commonly used scale in the world for everyday use. It was developed by Anders

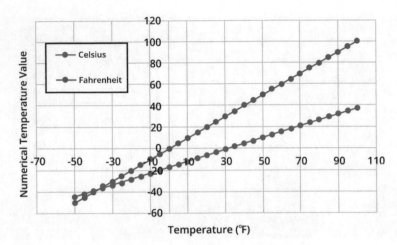

Fig. 15.2 Graph showing the relationship between the Celsius and Fahrenheit temperature scales

Celsius, who was a Swedish astronomer [3]. The scale is calibrated such that at one standard atmosphere of pressure the freezing point of water is 0 °C and the boiling point of water is 100 °C.

• **Kelvin (K)**: This scale was named after British engineer and physicist William Thompson, first Baron Kelvin [4]. The Kelvin scale uses absolute zero as its minimum point, but is defined based on universal constants (such as the speed of light) and therefore is not written with a degree sign. It is often associated with the Celsius scale because they have the same magnitude of one degree.

Both Fahrenheit and Celsius are used routinely in temperature measurements and displays, so it is good practice to be able to convert between the two unit systems. The next set of questions helps us to develop a method for converting between the Fahrenheit and Celsius scales.

15.2.1 **In the Celsius scale, what is the boiling point of water? What is the boiling point of water in the Fahrenheit scale?**

15.2.2 **In the Celsius scale, what is the freezing point of water? What is the freezing point of water in the Fahrenheit scale?**

15.2.3 **In Fig. 15.2, what type of relationship is the Celsius temperature scale, linear or parabolic?**

15.2.4 **In Fig. 15.2, what type of relationship is the Fahrenheit temperature scale, linear or parabolic?**

15.2.5 **What is the equation of a straight line?**

15.2.6 **In Fig. 15.2, what is the variable labeled on the x-axis?**

15.2.7 **Based on the previous question, for the equation of a straight line, what would be the y variable?**

15.2.8 **To make a linear equation two points are needed. Looking at Fig. 15.2, which two points would you use (hint: think back to questions 15.2.1 and 15.2.2).**

15.2.9 **Based on the previous questions, explain a procedure for converting from temperature in Fahrenheit to temperature in Celsius?**

15.2.10 **Based on the previous question, develop an equation for converting from temperature in Fahrenheit to temperature in Celsius.**

15.2.11 **Using the previous answer, develop an equation for converting from temperature in Celsius to temperature in Fahrenheit.**

15.2.12 **Based on Fig. 15.2, where do the Fahrenheit and the Celsius lines cross?**

15.2.13 **At what temperature do the Fahrenheit scale and the Celsius scale read the same value?**

As stated above, the Rankine and Kelvin scales are based on absolute zero or the absolute thermodynamic temperature. Most scientific constants and equations use temperature values in Rankine and Kelvin. So it is important to be able to covert from the displayed temperatures of Celsius and Fahrenheit to Rankine and Kelvin. The Rankine temperature is associated with Fahrenheit temperature and Kelvin is associated with Celsius. The conversion relationships are as follows:

$$T(°R) = T(°F) + 460$$
$$T(K) = T(°C) + 273$$
$$T(K) = 0.556 \times T(°R)$$

It is also important to understand about the relative difference in temperatures versus the absolute values displayed. This next inquiry helps us learn about absolute and relative temperatures.

15.2.14 **What is the equivalent value of $T = 85$ °F in Celsius, Rankine, and Kelvin?**

15.2.15 **On a winter day, in Syracuse, New York, the inside room temperature is maintained at 69 °F while the outdoor air temperature is a freezing 26 °F. Calculate the indoor-outdoor temperature difference in degree Fahrenheit, degree Rankine, degree Celsius, and Kelvin. First, convert each absolute value to the desired scale and then find the difference between the two.**

15.2.16 **Is a 1° temperature difference in Celsius equal to a 1° temperature difference in Kelvin, If so, why?**

15.2.17 **Is a 1° temperature difference in Fahrenheit equal to a 1° temperature difference in Rankine? If so, why?**

15.2.18 **Is a 1° temperature difference in Fahrenheit equal to a 1° temperature difference in Celsius? If so, why?**

15.3   Heat Transfer and Temperature Difference

The terms "heat" and "temperature" are very commonly used in engineering, science, and everyday life. While there is an interplay between the two concepts, they are not synonymous with each other. In the previous sections, we defined temperature as a single number that measures how much movement (kinetic energy) the molecules of a substance have. Heat, on the other hand, is a measure of how the total energy of an object is moving or changing. Temperature is a property an object can exhibit, heat is not. An object can lose or gain heat, it cannot have heat. It has a certain temperature. The symbol for heat is typically denoted as a capital Q. The next set of questions helps us compare and contrast heat and temperature.

15.3.1 **In a hot cup of tea, is it at a high temperature or a low temperature?**

15.3.2 **In a hot cup of tea, is the molecular motion of the water molecules high or low?**

15.3.3 **In a hot cup of tea, is there a high amount of energy or low amount of energy.**

15.3.4 **If a hot cup of tea is left out in a cold office, does the water heat up or cool down?**

15.3.5 **If a hot cup of tea is left out in a cold office, what happens to the tea water's temperature, molecular motion, and amount of energy?**

15.3.6 **Based on the previous questions, and describing the temperature and energy, explain what is happening to the cup of tea in the cold office.**

15.3.7 **Based on the previous answer, give a definition of heat transfer.**

15.3.8 **Based on temperature, which way does heat always travel?**

15.3.9 **For the hot cup of tea that is left out in a cold office, when will the heat transfer stop?**

15.3.10 **In your own words, explain the similarities between heat and temperature.**

15.3.11 **In your own words, explain three differences between heat and temperature.**

15.3.12 **Explain how heat and temperature interact with each other.**

Heat transfer is similar to the engineering term *Work* because it is a transfer of energy (we will discuss more about Work in Chap. 17). This can be seen in its measurement and units. Temperature is measured by a thermometer device with units previously discussed (Celsius, Kelvin, Fahrenheit, Rankine), whereas heat transfer is measured using a calorimeter and has units of energy. There are three common units used to quantify heat transfer, each with its own unique history and current usage. The following is a brief description of each and Table 15.1 are the main conversion standards used today. These standards were developed throughout history using experiments.

Table 15.1 Conversion factors for heat transfer and thermal energy

SI Unit	US Unit
1055 J =	1 Btu
4.186 J =	1 cal
252 cal =	1 Btu

- The SI unit system of thermal energy is the **Joule (J)**. The Joule went through a tumultuous history to become a standard unit [5], but around the early 1900s it became the standard SI unit of energy measurement. Since the SI systems make no distinction between the units of thermal energy and mechanical energy, the joule is defined in terms of fundamental dimensions (length, mass, time). It is defined as the energy transferred to an object when 1 N of force acts on that object through a distance of 1 m. Namely, $1 \text{ J} = 1 \text{ N m} = 1 \text{ kg m}^2/\text{s}^2$.
- The US unit of thermal energy is the **British thermal unit (Btu)**. The Btu is derived as the amount of heat required to raise the temperature of 1 pound mass of water by 1 °F.
- Another unit of thermal energy is the **calorie (cal)**. It is defined as the amount of heat required to raise the temperature of 1 g of water by 1 °C. In common use, the "calorie" that is used as a measure of nutrition is typically symbolized with a capital C, as **Cal**. This unit is actually a kilocalorie, namely, 1 Cal = 1000 cal.

In many cases, heat transfer is measured over time particularly in many heating and cooling operations. This is known as *Thermal Power*. The unit of thermal power for the SI system is known as the Watt (W), and it is derived as 1 J /s. The US unit of thermal power is the Btu/h. Although the Watt is a largely accepted unit for power in most places because of the prevalence of the SI unit system, the Btu/hour measure is still commonly used to describe the power output of many industrial applications including steam generators, home heating (furnaces), and air conditioning. The next set of questions lets us practice the conversion of thermal power units.

15.3.13 **Name three sources of heat (they can be natural or synthetic).**
15.3.14 **What is 1 W in J/s?**
15.3.15 **Using the units of energy and time, derive what 1 W equals in Btu/h.**
15.3.16 **If an application uses 2000 cal in 8 s, what is that equal in Watts?**
15.3.17 **How many Joules are in 1 Cal?**

As we just learned, the heat of an object is the cumulative energy of all the molecular movement inside that object. It is a form of energy that gets transferred from one object to another because of the differences in their temperature, and it moves from hotter sources to cooler ones. To measure heat, the units of energy are used (Btu, calories, or joules). The transfer of heat typically occurs in three distinct ways; Conduction, Convection, and Radiation. The next three sections look in detail at each of the different means of heat transfer.

15.4 Conduction

Of the three forms of heat transfer, conduction is typically the easiest to recognize, because it occurs through physical contact [6]. **Conduction** is heat transfer that occurs between molecules when they are in direct contact with each other. If two objects are at different temperatures and they are touching, the molecules in the hotter object are moving faster than the molecules in the cooler object. The faster moving molecules bump into the slower moving molecules at the surface of contact and transfer their energy, making the slower molecules speed up. This results in a rise in the temperature of the cooler object for as long as heat is still being added. Typically the objects through which conduction occurs are solid matter and stationary fluids. Typically gases have a lower thermal conductivity than liquids, and liquids have a lower thermal conductivity than solids.

Mathematically, conduction is modeled by **Fourier's Law**. It is named after Jean-Baptiste Joseph Fourier, the French mathematician and physicist who developed the principle [7]. The law quantifies the conduction heat transfer process by relating the *rate* of heat transfer to three key items: (1) temperature difference, (2) geometrical properties, and (3) material properties. Fourier's law can be stated in two equivalent mathematical forms. These equations are used to compute the amount of *energy* being transferred *per unit time*. The integral form looks at the amount of energy flowing into or out of a body as a whole. Students will study this form if they take an upper level heat transfer class. The differential form looks at the local temperature difference [8]. The simplest form of the differential equation assumes that the conditions are steady state, the heat being conducted in an object is flowing in only one direction, and that the material proprieties of the object are constant. For this case, Fourier's Law can be written to solve for the *heat rate* (denoted by q) by the following equation:

$$q = kA \frac{\Delta T}{L}$$

where A is the cross-section area normal to the heat flow, L is the thickness of the object, ΔT is the temperature difference, from hot temperature to cold temperature, across the material thickness, and k is a property of the material known as thermal conductivity. The next set of questions lets us explore conduction and Fourier's Law.

15.4.1 **Give three examples of conduction experienced in our everyday lives.**

15.4.2 **Rank these materials in order from highest to lowest thermal conductivity: water, Brass, and Helium.**

15.4.3 **Look at the equation for Fourier's Law, what are the fundamental dimensions and SI units for the variable L?**

15.4.4 **Look at the equation for Fourier's Law, what are the fundamental dimensions and SI units for the variable A?**

15.4.5 **Look at the equation for Fourier's Law, how can you expand the term ΔT to include the hot and cold temperatures that are driving the conduction?**

15.4.6 **For the term ΔT, what are the SI units?**

15.4.7 **The symbol q stands for heat rate, this is the amount of *energy* per unit *time*. Based on what you previously learned about energy units, what are the SI units for q?**

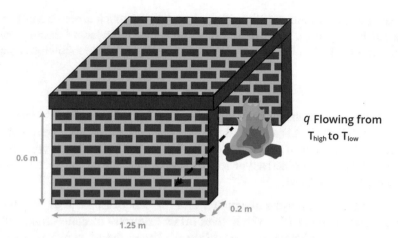

Fig. 15.3 One dimensional thermal conduction through a fireplace wall

15.4.8 **Look at the equation for Fourier's Law, based on the previous answers, use dimensional analysis to find the SI units of thermal conductivity.**

15.4.9 **A wall of a fireplace is shown in Fig. 15.3 is made of fire clay brick with thermal conductivity of 1.4 W/m•K. When the fire is burning, the temperature inside the fireplace was measured to be 500 °C and outside was 25 °C. With the dimensions of the fireplace shown in the figure, what is the rate of heat transfer through the 0.6 m × 1.25 m side of the wall?**

15.5 Convection

Heat transfer occurs by **Convection** when a surface comes into contact with a moving fluid and their temperatures are different. Energy transfer due to convection occurs because of a combination of two mechanisms; advection and diffusion. Advection refers to the directional movement of the bulk fluid, and diffusion accounts for random movement of particles along a concentration gradient [8]. There are two classifications for convective heat transfer, and they are distinguished according to the nature of the flow. Namely, what causes the fluid to flow. The first type is **free** convection (also known as *natural* convection). Free convection occurs because temperature variations in the fluid lead to natural density difference in the fluid, which makes the fluid move. For example, hot air rises and cold air falls. The second type is **forced** convection. Forced convection occurs because an external source (such as a fan or pump) is causing the fluid to move.

Mathematically, both free and forced convection can be modeled using **Newton's law of cooling**. It was developed by Sir Isaac Newton. The English mathematician, physicist, and astronomer whom the SI unit for force was named after [9]. Similar to Fourier's law, this rate equation relates heat transfer to a temperature difference, geometrical properties, and material properties. Newton's law of cooling can be written to solve for the *heat transfer rate* (denoted by q) by the following equation:

$$q = h \cdot A \cdot \left(T_{surface} - T_{fluid} \right)$$

where A is the area of the exposed surface and h is the heat transfer coefficient. The heat transfer coefficient is a material property that depends on the type of fluid and the physical situation in which the convection occurs. Therefore, typically heat transfer coefficients are derived or found experimentally for each specific system that needs to be analyzed.

15.5.1 **Give three examples of conduction experienced in our everyday lives.**
15.5.2 **Look at Table 15.2 which gives other examples of convection. Fill out the table, labeling each situation as either *forced* convection or *free* convection.**
15.5.3 **After a long run or workout, when you are trying to cool down is it better to sit in a room where the air is still or do sit in front of a fan?**
15.5.4 **Based on the previous question, is the value of h higher for free or forced convection?**
15.5.5 **When the air temperature outside is 70 °F, is that considered warm or cold? (Think: would you wear a T-shirt or a jacket to go outside).**
15.5.6 **When water temperature is 70 °F, is that considered warm or cold? (Think: the average temperature for residential swimming pools is between 78 and 82 °F).**
15.5.7 **Based on the previous two questions, if you are standing in 70 °F air then 70 °F water, which removes heat faster from your body?**
15.5.8 **Based on the previous three questions, which is the heat transfer coefficient higher, for liquids or gases?**
15.5.9 **A computer chip is exposed to a cooling fan and has a heat transfer coefficient of 30 W/m² •K. The exposed surface area is 10 cm². If the surface temperature of the chip is 30 °C and the temperature of the surrounding air is 20 °C, determine the heat transfer rate.**

Table 15.2 Situations where convection occurs

Situations where Convection occurs	Type of Convection
Blowing on pizza to cool it down before eating	
Setting a pan of freshly baked brownies on the counter to cool	
Turning on a ceiling fan on a hot day	
The ocean breeze	
Cumulus clouds forming in the sky	
Using a hair dryer	
Heat sink in a computer	

15.6 Radiation

When you sit by a campfire to stay warm you are experiencing the third type of heat transfer [8]. **Radiation** is the emission of thermal energy in the form of waves or particles from an object with high temperature to one with low temperature. And unlike conduction, the bodies are not in direct physical contact with each other. And unlike convection, radiation does not need a material medium to occur. It can occur in a vacuum. All matter emits radiation because atoms and molecules of an object always have vibrational and rotational movement, even if it is on a nano-scopic scale. The strength of the thermal energy depends upon the temperature of the object and its surface structure.

For a surface-emitting radiation to a much larger surface, the heat energy (q) from the surface is governed by the **Stefan-Boltzmann equation** which states:

$$q = \varepsilon \sigma A T_s^4$$

where A is the area of the surface, T_s is the temperature of the surface of the body, and σ is a constant of proportionality called the Stefan-Boltsmann constant ($\sigma = 5.67 \times 10^{-8}$ W/m^2 K or 1.714×10^{-9} BTU/h ft^2 R). The term ε represents emissivity. Emissivity is a surface property. It characterizes how efficiently a surface emits energy as compared to a blackbody. A *blackbody* is a theoretical heat transfer surface that perfectly absorbs all incoming radiation and reflects none back. The values of emissivity range from 0 to 1. A zero value of ε means the surface behaves like a shiny mirror and reflects all radiate energy. A ε value of 1 means the surface behaves like a perfect blackbody. Most surfaces are between $0 \leq \varepsilon \leq 1$ and are considered "gray" because they both absorb and reflect energy. Because most objects both absorb and reflect energy, radiation calculations become quickly complicated and require an in-depth understanding of the geometry and material of interest. Students will learn more complex radiation calculations in higher level engineering courses such as Heat Transfer. The next set of questions will give a brief introduction into the concept of radiation.

15.6.1 **Give three examples of radiation experienced in our everyday lives.**

15.6.2 **For the Stefan-Boltsmann equation, what are the SI and US units of symbol A?**

15.6.3 **For the Stefan-Boltsmann equation, what are the SI and US units of symbol ε?**

15.6.4 **For the Stefan-Boltsmann equation, what are the SI and US units of symbol T_s?**

15.6.5 **On a hot summer day, an asphalt driveway was sitting in the sun all day and reached a temperature of 135 °F. Determine the amount of thermal radiation energy that is radiating from the surface of a two car driveway that is 24 ft wide and 25 ft long. The emissivity of the driveway is 0.9.**

15.6.6 **On a hot summer day, if the same asphalt driveway as the previous questions was now covered all day, so it reached a temperature of 85 °F. Determine the percent difference in the amount of thermal radiation between the driveway being covered versus uncovered.**

15.7 Temperature-Related Material Properties

Many physical and material properties are a function of temperature. As a solid, liquid, or gas gets colder or hotter it will behave differently. For example, you know that hot air rises. This is because as air temperature increases, its density decreases. Three important temperature-related material properties that engineers regularly encounter are thermal resistance, thermal expansion, and specific heat.

1. **Thermal Resistance (R-value):** Thermal resistance is a measurement that determines the heat insulation property of a material or assembly of materials. It describes how easily the object loses heat. For example, a double paned window compared to a single pane window. To obtain an equation for thermal resistance, Fourier's law of conduction needs to be rearranged. When the equation for heat transfer (q) is arranged in terms of the ratio of temperature difference to thermal resistance it looks as follows:

$$q = kA\frac{\Delta T}{L} = \frac{\Delta T}{\dfrac{L}{kA}} = \frac{\text{Temperature difference}}{\text{Thermal resistance}}$$

where the term in the denominator is known as *thermal resistance for a unit area*, denoted as $R' = L/kA$. The SI units for R' are either °C/W or K/W and the US units are °F h/Btu or °R h/Btu. When you look at thermal resistance per unit area of a material it is referred to as the R-value or R-factor. It is denoted as $R = R' \cdot A = \dfrac{L}{k}$.

R-factors are used regularly in the construction industry to label things like insulation, brick, windows, and other construction building materials. These next few questions let us examine thermal resistance and R-values.

15.7.1 **From Fourier's equation given above, explain the relationship between heat transfer (q) and temperature difference.**

15.7.2 **From the equation given, explain the relationship between heat transfer (q) and thermal resistance.**

15.7.3 **From knowing the units for R', what are the SI and US units for F-factor?**

15.7.4 **The window of single pane glass show in Fig. 15.4a has a thickness of 7 mm, a width of 2 m, and a height of 1 m, calculate the thermal resistance and the R-value for this window.**

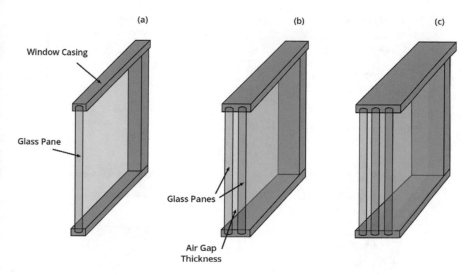

Fig. 15.4 (**a**) Single, (**b**) Double, and (**c**) Triple Pane Glass Windows. The thermal conductivity of air is 0.0263 W/m•K and the thermal conductivity of glass is typically 1.4 W/m•K

15.7.5 As we learned before, thermal conductivity is the measure of a material's ability to conduct heat. Based on the previous information, explain the relationship between thermal conductivity and thermal resistance.

Calculating thermal resistance and R-value for single objects is straightforward, but what happens when you have two materials layered together? Or a gas enclosed by a solid material? When calculating the thermal resistance and R-value of multiple layers of materials or substances it is convenient to use the *thermal resistance circuit analogy*. This technique equates the flow of thermal energy to the flow of electricity through wires. This helps simplify the analysis of a complicated heat transfer system and makes it easier to visualize. The analogy is based on Ohm's law, which is a formula that relates the voltage, current, and resistance in an electrical circuit. Ohm's law states that the voltage (V) through a conductor is directly proportional to the current (I) that flows through it. When written in equation form is means that voltage is equal to the current flowing through the wire times the electrical resistance ($R_{electric}$) that wire has or $V = I·R_{electric}$. (We will discuss this concept in more detail in Chap. 16.) Then if we rearrange Ohm's law in terms of current and compare it to the thermal resistance definition we can see the similarities, shown in Fig. 15.5. Then just like circuits, where the electrical resistances of different components are summed together, the thermal resistances of different materials can also be combined together to develop a composite thermal resistance value. The next set of questions helps us to calculate the composite thermal resistance of a system of different materials.

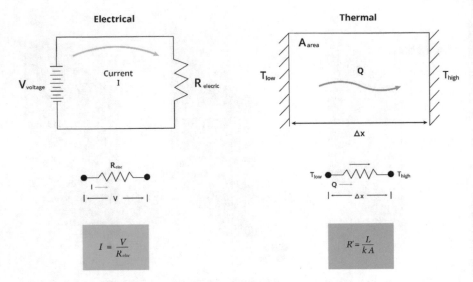

Fig. 15.5 Comparison between Ohm's Law and thermal resistance

15.7.6 **If the dimensions from Question 15.7.4 are the same for the windows in Fig. 15.4b and the air gap between each pane is 10 mm. Develop equations for the thermal resistance and the R-value for this window.**

15.7.7 **If the dimensions from Question 15.7.4 are the same for the windows in Fig. 15.4b and the air gap between each pane is 10 mm. Calculate the thermal resistance and the R-value for this window.**

15.7.8 **If the dimensions from Question 15.7.4 are the same for the windows in Fig. 15.4c and the air gap between each pane is 10 mm. Develop equations for the thermal resistance and the R-value for this window.**

15.7.9 **If the dimensions from Question 15.7.4 are the same for the windows in Fig. 15.4c and the air gap between each pane is 10 mm. Calculate the thermal resistance and the R-value for this window.**

15.7.10 **Based on the Question 15.7.4 and questions 15.7.7 and 15.7.10, which window would keep a house the warmest? Explain why?**

15.7.11 **Perform the same calculation for question 15.7.7 if the gap was now filled with Argon gas (thermal conductivity of 0.016 W/m K.) Calculate the R-value. How have the insulative properties of the windows changed now? (Note: Window manufacturers typically fill the air gaps with Argon or Krypton gas.)**

15.7.12 **If you had a double paned glass filled with argon in the gap, and dimensions the same as before (thickness of 7 mm, a width of 2 m, and a height of 1 m, and an air gap = 10 mm), if the outside air was 3 °C and the inside air was 22 °C, calculate the heat transfer rate through the window.**

2. **Thermal expansion**: This phenomenon is found prevalently in engineering and everyday life. Holes that occur in sidewalks and roads after the winter season are a function of the thermal expansion and contraction of water in cracks as the temperatures rise and fall. Thermal expansion is the response of a phase of matter to change its shape, area, or volume when a change in temperature occurs. It is typically expressed as a fractional change in length or volume per unit temperature change. For the expansion of a solid material, the coefficient of linear expansion (α_L) is used. The coefficient of linear expansion is defined as the ratio of the change in length over the original length times the change in temperature. For an expanding liquid or gas, the coefficient of volumetric expansion (α_V) is employed. It is defined as the change in volume per the original volume times the change in temperature. For homogenous solid materials, the coefficient of volumetric expansion is three times the coefficient of linear expansion, namely $\alpha_V = 3\alpha_L$. This next inquiry explores thermal expansion coefficients.

15.7.13 **Give three examples of thermal expansion experienced in everyday life.**

15.7.14 **Given three examples of thermal expansion experienced in engineering applications.**

15.7.15 **Based on the written definition above, write the mathematical expression for the coefficient of linear expansion.**

15.7.16 **Based on the previous question, what are the SI and US units of coefficient of linear expansion?**

15.7.17 **Based on the written definition above, write the mathematical expression for the coefficient of volumetric expansion.**

15.7.18 **Based on the previous question, what are the SI and US units of coefficient of volumetric expansion?**

15.7.19 **Calculate the change in length of a 2000 ft long stainless steel I-beam if it changes temperature by 100 °F (note: α_L for stainless steel is $2.9 \times 10^{-6}/°F$.)**

3. **Specific heat**: Have you ever gone to the beach and stepped on the sand and it burned your feet, only to run down to the ocean which was nice and cool? Why does this happen? Both the ocean water and the sand are exposed to the same amount of the sun's thermal energy throughout the day. Due to various structural factors, different substances absorb heat at different rates. In the eighteenth century, Scottish medical doctor and Professor Joseph Black observed this concept and quantified it as *specific heat* [10]. Denoted by the symbol c, it is the heat capacity of a sample of substance divided by the mass of the sample. In general, it is the amount of heat energy that must be added, to one unit of mass of a substance in order to cause an increase of one unit of temperature. In the SI unit system, specific heat is defined as the thermal energy required to raise the temperature of 1 kg mass by 1 °C. In the US unit system, specific heat is defined as the thermal energy required to raise the temperature of 1 lb mass by 1 °F. The specific heat equation quantifies the relationship between the required thermal energy ($E_{thermal}$), mass of the object (m), its specific heat (c), and the temperature change (ΔT) and it is given by

$$E_{thermal} = mc\Delta T$$

The specific heat relationship does not apply if a phase change is encountered, because the heat added or removed during a phase change does not change the temperature. The next set of questions helps introduce us to the concept of specific heat and demonstrates the use of the equation.

15.7.20 **What are the SI units for thermal energy? (Hint: look back at previous sections if necessary).**

15.7.21 **What are the SI units for mass? (Hint: look back at previous sections if necessary).**

15.7.22 **What are the SI units for temperature? (Hint: look back at previous sections if necessary).**

15.7.23 **Based on the previous questions and the given equation, what are the SI units for specific heat?**

15.7.24 **What are the US units for specific heat?**

15.7.25 **Using Table 15.3, if you exposed water and sand to the same amount of thermal energy from the sun, which one would feel hotter to the touch?**

15.7.26 **Using Table 15.3, if you exposed water and copper to the same amount of thermal energy, which one would feel hotter to the touch?**

15.7.27 **The following materials start out at 20 °C and are exposed to a heat source that put of 150 J every second: 1 kg lead, 1 kg sand, and 1 kg liquid water. Which material will have the greatest temperature rise after 10 s? Which material will have the smallest temperature rise after 10 s?**

15.7.28 **Based on the previous questions, write a statement relating a material's specific heat (c), its ability to absorb heat, and its temperature rise.**

15.7.29 **A copper disk with a diameter of 20 cm and thickness and 5 cm is exposed to a heat source of 300 J every second. The density of the copper is 8960 kg/m³. If no heat is lost to its surroundings, what is the rise in temperature the disk will experience after 30 s?**

15.7.30 **Can the following statement be true: $c_{solids} < c_{liquids} < c_{vapors}$? Why or why not?**

Table 15.3 Specific heat of some materials at a constant pressure

Material	Specific heat (c)
Air	1007
Copper	385
Lead	129
Sand	800
Liquid water	4180
Hydrogen gas	14,300

End of Chapter Questions

IBL Questions

IBL1: You want to buy a stove with a front facing area of 0.5 m high and 1.5 m long and 0.3 m thick. You are trying to decide between a cast-iron or a wrought iron stove. The cast-iron stove has a heat transfer rate of 74,750 W when the temperature is 600 °C inside the stove while the outside is typically 25 °C. The wrought iron stove has a heat transfer rate of 77,438 W when the temperature is 550 °C inside the stove while the outside is typically 25 °C. Answer the following questions:

1. If it is assumed the heat only flows in one direction (out of the stove), is the condition of the stove steady state?
2. What equation would be used to solve for the heat rate of the stove?
3. What is the area over which the heat will travel?
4. What is the length over which the heat will travel?
5. What is the change in temperature that drives the heat transfer for both materials?
6. What is the driving variable that effects the heat transfer rate?
7. Explain a process for deciding which type of material to choose for the stove.
8. Calculate the thermal conductivity for both potential materials.
9. Which material would be chosen for use in the stove? Explain why.

Practice Problems

1. Calculate the equivalent temperature of $T = 200$ °F in units of Rankine, Celsius, and Kelvin.
2. Rate the following materials from highest to lowest R-value. Note thermal conductivities:

   • Layer of human fat that is 2 cm thick
   • Layer of human muscle that is 4 cm thick
   • Layer of human skin that is 2 mm think

References

1. Fahrenheit temperature scale, Encyclopædia Britannica Online. Retrieved 14 December 2020.
2. "Rankine". Merriam-Webster Dictionary Online. Retrieved 15 December 2020.
3. Celsius temperature scale, Encyclopædia Britannica Online. Retrieved 15 December 2020.
4. Lord Kelvin, William (October 1848). "On an Absolute Thermometric Scale". Philosophical Magazine. Archived from the original on 1 February 2008. Retrieved 15 December 2020. https://web.archive.org/web/20080201095927/http://zapatopi.net/kelvin/papers/on_an_absolute_thermometric_scale.html

5. Gregory S. Girolami, A Brief History of Thermodynamics, As Illustrated by Books and People J. Chem. Eng. Data 2020, 65, 2, 298–311, Publication Date: September 12, 2019. https://doi.org/10.1021/acs.jced.9b00515

6. J.M.K.C. Donev et al. (2020). Energy Education - Thermal conduction [Online]. Available: https://energyeducation.ca/encyclopedia/Thermal_conduction#cite_note-1.   [Accessed: January 4, 2021].

7. Fourier, Joseph. (1878). The Analytical Theory of Heat. Cambridge University Press (reissued by Cambridge University Press, 2009; ISBN 978-1-108-00178-6)

8. Bergman, T.L., Lavine, A.S. Incropera, F.P., Dewitt, D.P., (2011) Introduction to Heat Transfer (6th Edition), John Wiley & Sons, Inc. ISBN-13: 978-0470-50196-2.

9. Westfall, Richard S. (1983) [1980]. Never at Rest: A Biography of Isaac Newton. Cambridge: Cambridge University Press. pp. 530–531. ISBN 978-0-521-27435-7.

10. Laidler, Keith, J. (1993). The World of Physical Chemistry. Oxford University Press. ISBN 0-19-855919-4.

Chapter 16
Electric Current

Abstract One of the most powerful and useful forms of energy is electricity. It is so valuable that there is an entire engineering discipline dedicated to this field of study; Electrical Engineering. Even though electrical engineers learn about designing, analyzing, and controlling the fields of electricity, magnetism, and power generation, it is beneficial for *all* disciplines of engineering to learn the fundamentals. The objective of this chapter is to introduce the concepts of electricity and circuits. To begin, we will review concepts of electricity that may have been learned in physics class, then we will discuss measurements of charge and current. Also, we will discuss mechanisms of circuits such as voltage, potential, and resistance. We will examine and compare the two major types of current (DC vs. AC current). Finally, we will provide an introduction to simple circuit theory, including Ohm's Law, Kirchhoff's laws, and analysis of parallel and series circuits.
By the end of this chapter, students will learn to:

- Define current and explain the importance of current in everyday life and in engineering analysis.
- List examples of current and related variables.
- Recognize and explain the units of current, voltage, and potential.
- Compare and contrast DC and AC current.
- Calculate RMS voltage current.
- Explain why AC current is dominantly used for power transmission.
- Explain what a resistor and capacitor are.
- Identify a resistor value based on the color bands.
- Identify nomenclature and symbols for simple electric components.
- Explain how simple circuits work.
- Apply Kirchhoff's law to simple resistive circuits.
- Apply Ohm's law to simple electrical situations.
- Calculate electric power consumption of simple electrical components.

© Springer Nature Switzerland AG 2022
M. Blum, *An Inquiry-Based Introduction to Engineering*,
https://doi.org/10.1007/978-3-030-91471-4_16

16.1   Importance of Electric Current and Parameters

The concept of electricity can sometimes be challenging to grasp because we cannot directly visualize many of the parameters. For example, we can see a lamp light up, but with our naked eye we cannot see the energy flowing through a wire when the light switch gets flicked. We learned from Chap. 13 that all matter is made of molecules or atoms. Molecules are the smallest particle that a substance can be broken down to and retain all its characteristics. Atoms make up molecules, and they contain smaller particles called electrons, protons, and neutrons. The protons and neutrons reside in the atom's nucleus, and the electrons travel in an orbit around the nucleus. Neutrons have no charge, but significant mass. Protons have significant mass and a positive charge. Electrons have extremely small mass and a charge that is equal in numeric magnitude to a proton, but opposite in sign.

Atomic structures increase in complexity as the number of protons and electrons increase. Simple atoms, such as helium, only have two protons in the nucleus and two electrons in two orbits, also called orbital shells. The shells close to the nucleus can only contain a maximum of two electrons, but that number increases as the shells get farther away. When there are more orbital electrons and the shells are not completely full, atoms can combine together by having orbital electrons share shells. This creates molecules. How closely the atoms and molecules are packed together dictates the state of a substance (solid, liquid, vapor, etc.).

Electrons are prevented from leaving the atom by an electric force of attraction. Recall back to Chap. 14, when we talked about forces. It was discovered by French military engineering and physicist Charles-Augustin de Coulomb that an electric force exists between any object with a charge [1]. Coulomb's Law follows that electric force is inversely proportional to the square of the distance between the objects and directly proportional to the magnitude of each charge. Mathematically it is written as

$$F_{12} = \frac{k_e q_1 q_2}{r^2}$$

where $k_e = 8.99 \times 10^9$ N m^2/C^2 is a proportionality factor known and electric force constant or electrostatic constant [2]. The charge magnitude of each object is labeled as q_1 and q_2 and r is the distance between the objects. Since protons and electrons have equal and opposite charges, when there is an even number, the force keeps the electrons in place. When the charges are unbalanced, electrons can drift around, and just like energy, electric charge is not created or destroyed, so it can be transferred from one object to another. Finally, both electrical theory and experiments have shown that when the charges q_1 and q_2 have the same sign, the force is repulsive. Then when the charges q_1 and q_2 have the opposite sign, the force is attractive. This all leads to the concept of electric current. The next set of questions help us define current, its importance, and other related parameters.

16.1.1. **Think back to Chap. 13 and mass flow rate, for water to flow through a pipe, does it flow from a region of high pressure to low pressure, or from a region of low pressure to high pressure?**

16.1.2. **Think back to Chap. 15 and heat flow, does thermal energy flow from a region of high temperature to low temperature, or from a region of low temperature to high temperature?**

16.1.3. **Based on the previous two questions, if there was an electric charge difference between two objects (also known as electric potential), will electric charge flow from the higher electric potential to the lower potential region or from the lower electric potential to the higher potential region?**

16.1.4. **Based on question 16.1.1, how does the water flow in a confined and controlled manner?**

16.1.5. **Based on the previous question, how do you think electric charge flows in a confined and controlled manner?**

16.1.6. **Based on all the previous questions, explain in your own words what** *current* **is?**

16.1.7. **Give three examples of materials that are considered good conductors.**

16.1.8. **In your own words, explain what a conductor is.**

16.1.9. **Give three examples of materials that are considered good insulators.**

16.1.10. **In your own words, explain what an insulator is.**

16.1.11. **In your own words, explain why current is important in everyday life and engineering.**

16.2 Measurements of Current, Voltage, and Potential

As we discovered in the previous section, an electric current is a stream of charged particles through an electrically conducting substance. The SI and US unit of *charge* is **Coulomb (C)**, named after Charles-Augustin de Coulomb and defined as the amount of charge that passes through an electrical conductor in one section, when it is carrying a current of one ampere. Typically the particles (also known as "charge carrier") are electrons, but sometimes they can also be ions. **Current** measures the net rate of flow of electric charge past a certain region [3]. Often in electric circuits, the charge carriers are electrons and they move through metal wires. The unit of measurement for current (in both the US and SI unit systems) is the **ampere**, abbreviate **amp** with symbol **A**. One **ampere (A)** is defined as the flow of 1 unit of charge per second or one ampere equals one Coulomb per second.

The amount of current that flows through an electrical element depends on the resistance the element has to the flow of charge and the voltage available across the element. **Voltage**, also known as electrical potential, is the amount of work required to move the electric charge, or current. Equating electrical circuits back to water pipes, think about voltage as the pressure that comes from a power source (such as a battery) that pushes the current (charged electrons) through the wires. Voltage is measured in units of energy per unit charge in units of **Volts (V)**. The next set of questions helps us explore units of current, voltage and perform calculations on various electrical units.

16.2.1. **If a coffee maker has 2 C of charge flowing through the heating element each second, how many amperes does it draw?**

16.2.2. **What are the SI units of energy (recall back to Chap. 15)? What are the SI base units?**

16.2.3. **What are the fundamental SI units of Ampere?**

16.2.4. **Based on the previous two questions, what are the base SI units of the unit Volts?**

16.2.5. **Based on the previous question, what are the fundamental dimensions of the unit Volts?**

16.2.6. **Using the previous two questions and think back to Chap. 15 on Temperature and the definition of a Watt (W). Develop the relationship between Watts, Volts, and Amps.**

16.2.7. **In the bathroom, you have a 15 amp GFI (Ground-Fault Circuit Interrupter). This means that if you use more current than the wire can handle, the breaker will trip and the circuit will shut off. You plug in a 2000 W hair dryer, the primary voltage supply in a domestic residence in the United States is 120 V. Knowing this, will you trip the breaker if you turn on the hair dryer?**

16.3 DC and AC Current

When thunder and lightning strike it is not just an energy exchange happening from the clouds to the earth, but a reaction in the air to the energy passing through it. To detect this energy transfer, we must use measurement tools such as multimeters, spectrum analyzers, and oscilloscopes to visualize what is happening with the charge in a system.

A battery is a constant voltage source, meaning that once the current is established (for example, like connecting the battery to a wire) the current will also be constant. If the flow of electric charge is constant in only one direction it is called **Direct current (DC)**. Direct current was invented by Italian physicist Alessandro Volta [4] but was brought to the mainstream in the United States by Thomas Edison in the 1880s with his invention of the incandescent bulb [5]. Anything that runs off a battery relies on DC for power, and almost all home electronics run off of DC power now because they have either a rectifier or use a USB cable. Since we cannot view what is happening with the charge in a system with the naked eye, we must use measurement tools such as multimeters, spectrum analyzers, or oscilloscopes. Figure 16.1a is a DC current graph that would be visualized off an oscilloscope and it shows the voltage and the current (I_{DC}) are constant over time. The graph shows that for most simple circuits' problems we can assume that most DC sources to

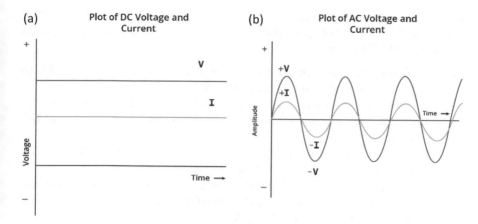

Fig. 16.1 Plots of voltage and current over time for (**a**) Direct current and (**b**) Alternating C voltage

provide a constant voltage over time. However, in reality, a battery will slowly lose its charge over time, meaning that the voltage will drop as the battery is used.

During the early years of electricity, the DC current was easily applied in US cities. However, since it is constant, it is not easily converted to a higher or lower voltage. This meant that power plants needed to be located within 1 mile of the end users, which made it difficult to transmit DC current over long distances to rural areas. Thus, competition to DC current arose from rival engineers George Westinghouse and Nikola Tesla, who advocated electricity transfer by **Alternating current (AC)** [5]. Alternating current is when the flow of electric charge periodically changes direction, and as a result, the voltage level also reverses. The cyclic manner of the current allows the voltage to easily be increased or decreased, so it can easily and cost effectively travel over long distances. The electricity that powers our buildings and homes is predominantly powered by alternating current today.

The most common type of AC is the sine wave (Fig. 16.1b). Therefore, the AC waveform can mathematically be described using a common sine wave equation; $V(t) = V_0 \sin (2\pi f t)$. Where $V(t)$ is the voltage as a function of time. Time (t) is typically measured in seconds, and as an independent variable, as it changes, the waveform varies. The amplitude is represented by V_0, meaning this is the maximum and minimum voltage the wave travels in either direction. The variable f is the frequency of the sine wave measured in hertz (cycles per second) and it tells how many times a particular cycle (a rise and a fall) of the sine wave occurs within 1 s. The constant 2π converts the frequency from hertz to angular frequency (radians per second).

Most often, engineers are concerned with average voltage or current rather than its fluctuations but looking at Fig. 16.1b it can be seen that a simple average calculation will not work. So the **RMS average** current and voltage are used. RMS stands for *Root Mean Square*. This particular kind of average is obtained by first squaring the quantities of interest, finding the mean of those quantities, and then taking the square root. For a sinusoidal AC voltage, this equation is $V_{RMS} = V_0 / \sqrt{2}$. Another way to think about RMS voltage is that it is the amount of AC power that produces the same effect as an equivalent DC power. Most residential electrical devices are labeled with the RMS voltage or power. The next set of questions helps us learn about DC and AC current.

16.3.1. **What is the power that comes through to your house, out of your home outlets, from the electric company; AC or DC?**

16.3.2. **Give three examples of everyday items that run off of DC power.**

16.3.3. **In your own words, explain the main differences between AC and DC voltage.**

16.3.4. **If Fig. 16.1a showed an AA battery that typically provides 1.5 V, write a mathematical equation for the DC voltage over time $V(t)$.**

16.3.5. **AC can come in several forms, as long as the voltage and current are alternating. Draw an example of graphs of an AC wave if it was (1) a square wave, (2) a triangle wave.**

16.3.6. **Using Fig. 16.1b, calculate the arithmetic mean (average) voltage of the sine wave if the voltage varies from 150 to −150 V.**

16.3.7. **Based on the previous question, explain why an RMS average needs to be used for AC voltage instead of a typical arithmetic mean.**

16.3.8. **In the United States, the power provided to our homes is AC voltage that is an average 120 V. What is the maximum and minimum voltage for this average?**

16.3.9. **Based on the previous question, what is the amplitude of the power provided to our homes by AC current?**

16.3.10. **If the AC current comes from US homes with a frequency of 60 Hz, write an equation for the sinusoidal wave function.**

16.3.11. **Based on the equation you just developed, fill in the blanks in the given sentence: *The equation shows that the AC voltage swings from _____ to _____ and back _____ times every second.***

16.3.12. **For the equation you just developed, what is the equivalent DC voltage?**

16.4   Resistors and Capacitors

In the previous Sects. 16.2 and 16.3, we started to explore the basics of electronics with the concepts of current and voltage. Another very important basic concept is resistance. *Electrical Resistance* is material or substance's ability to resist current flowing through it. **Resistance (R)** is measured in Ohms (Ω). The unit was named after German physicist and mathematician Georg Simon Ohm who pioneered many of the fundamentals of electricity [6]. In simple circuits, there are electrical components called resistors. Resistors are used to control the flow of current or split voltages in a circuit, so they are often installed to protect sensitive components. They can resist the flow of DC or AC current and are one of the most widely used components within electronic equipment.

Resistors come in two broad groups: fixed and variable. Both types are commonly used in electronic circuits. Below are some details on each kind and when to choose one type versus the other:

- **Fixed-value Resistors** supply a constant, factory set resistance. These are used when the voltage needs to be divided in a particular fashion, or if the flow of current needs to be restricted within a certain range.
- **Variable-value Resistors**, also called potentiometers, allows the resistance to be adjusted from essentially zero ohms to a factory set maximum value. Variable resistors are used when the amount of current or voltage supplied to the circuit needs to be varied.

Depending on the application, resistors are available in many different shapes, sizes, and values. General purpose resistors include wirewound, surface mount, carbon film, and metal film. Most leaded resistors with a standard power rating have a pattern of colored bands. The color codes are used to indicate the resistance value, tolerance of the device, and sometimes even the temperature coefficient rating. The number of color bands can range from three up to six, however four bands are the most common occurrence. The directions to read resistor values are below and refer to Fig. 16.2.

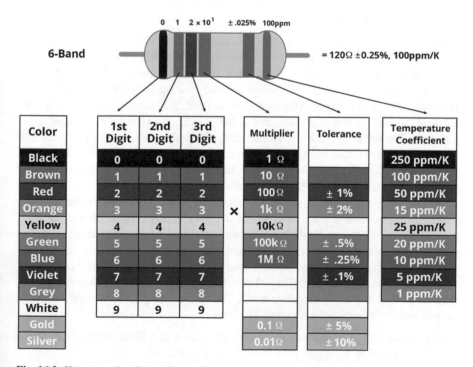

Color	1st Digit	2nd Digit	3rd Digit	Multiplier	Tolerance	Temperature Coefficient
Black	0	0	0	1 Ω		250 ppm/K
Brown	1	1	1	10 Ω		100 ppm/K
Red	2	2	2	100 Ω	± 1%	50 ppm/K
Orange	3	3	3	1k Ω	± 2%	15 ppm/K
Yellow	4	4	4	10k Ω		25 ppm/K
Green	5	5	5	100k Ω	± .5%	20 ppm/K
Blue	6	6	6	1M Ω	± .25%	10 ppm/K
Violet	7	7	7		± .1%	5 ppm/K
Grey	8	8	8			1 ppm/K
White	9	9	9			
Gold				0.1 Ω	± 5%	
Silver				0.01 Ω	±10%	

Fig. 16.2 How to read resistor color codes

- **Read the resistor starting from the left side and then over each band to the right**. Determine which is the left and the right side by looking for (1) Groups— many resistors have colored bands grouped together; (2) Resistors never start with a metallic band on the left. The metallic bands always denote the multiplier and/or the tolerance.
- **Reading from left to right, the first few bands denote the resistance value in ohms**. On a *three* or *four* band resistor the first *two* bands indicate resistance value. On a *five* or *six* band resistor the first *three* bands indicate resistance value.
- **The next band after the value represents the multiplier**. On a *three* or *four* band resistor it is the third band. On a *five* or *six* band resistor it is the fourth band. The multiplier shifts the decimal place to change the magnitude of the value, starting as little as milliohms up to megaohms.
- **The next band after the multiplier is the tolerance**. The fourth color band signifies tolerance. The tolerance numbers represent how far off from the nominal value the actual resistance could deviate. Note that if you are looking at a three band resistor this band will be absent. That means there is a default tolerance of ±20%.
- **Higher precision resistors are five and six bands**. A five band resistor has three resistor value bands. A six band resistor is a five band type with an additional ring after the tolerance. This typically indicates the temperature coefficient (ppm/K) specification. For example looking at the most common sixth band color, brown, this signifies that every temperature change of 10 °C changes the resistance value by 0.1%.

Another common electrical component in circuits is a capacitor. A **capacitor** has oppositely charged electrodes with a dielectric material inserted between the electrodes so that it can store electrical charge. This allows them to be used in many applications for safety such as filters and memory storage to protect electronic components from power surges. They can also be used in electronics that need timing elements such as strobe lights and car windshield wipers. The unit for a capacitor is the farad. The **farad (F)** unit was named after Michael Faraday, an English scientist who was another leader in the study of electricity and electromagnetism [7]. One farad is equal to one coulomb per volt. A farad is a large unit; therefore, capacitor sizes are commonly expressed in mico-farad or pico-farads.

When designing a circuit, distinct symbols can be used for various components. The simplest symbol is for a wire, which is typically represented by a solid black line. There are different symbols for the various types of resistors used. Also, the circuit symbols used in the USA, set by the American National Standards Institute (ANSI), are different than the symbols used within Europe, set by the International Electrotechnical Commission (IEC). Table 16.1 shows figures of the various components as well as various circuit symbols. The next set of questions helps us to look at resistors and capacitors and circuit symbols.

16.4.1. **Which type of resistor do you think is used in a light switch that allows you to dim the lights; fixed-value or variable resistor?**

Table 16.1 Examples and symbols of common circuit components

Component	Figure	USA Circuit Symbol	European Circuit Symbol
Wire			
Resistor (fixed)			
Capacitor			
Battery			
Switch			

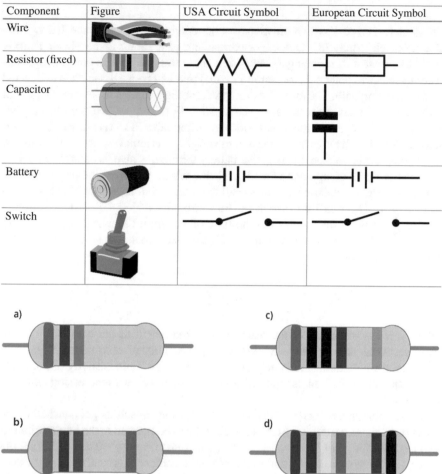

Fig. 16.3 Examples of various resistors and band codes

16.4.2. **Which type of resistor do you think is used in a light switch that allows you to adjust the volume on your car radio; fixed-value or variable resistor?**

16.4.3. **Which type of resistor do you think is used in LED lights; fixed-value or variable resistor?**

16.4.4. **Explain the main difference between an IEC resistor symbol and the ANSI symbol. Why do you think this is?**

16.4.5. **How many farads is a microfarad? How many farads is a picofarad?**

16.4.6. **For each of the resistors shown in Fig. 16.3, calculate the resistance and tolerance values.**

16.5   Simple Circuits and Circuit Power

An **electrical circuit** is a combination of conducting components connected together that move electricity [8]. One major component of a circuit is the **power source**, with the simplest forms being a battery or a home wall outlet. The power source has a certain electric potential (measured in volts) that can be used once it is connected to a conducting path. The electricity moves from high voltage to lower voltage. For DC voltage sources, there are always two sides. For example, a battery has a positive side and a negative side. These sides are often labeled as **terminals**. The positive terminal has the higher voltage and the negative terminal has the lower voltage. When measuring voltage, the negative battery terminal is always considered 0 V.

Another major component of a circuit is the **wires**. Wires are the physical channels that carry the electric current. Recall from the previous section that resistance is a measure of how easily a current can flow through a material. Resistance through a wire depends on four things (1) the type of material, (2) wire length, (3) wire diameter, and (4) operating temperature. Wire resistance (R) can be measured by the following equation

$$R = \frac{\rho L}{A}$$

where ρ is the resistivity of the material, L is the length of the wire and A is the cross-sectional area of the wire. Resistivity has units of ohm-meters. Typically wires are made with materials that have low resistivity. The resistivity of materials vary with temperature. In general, the resistance increases with increasing temperature [9].

Other components such as switches, resistors, and capacitors get attached to the wires throughout the circuit. An electric circuit can be equated to the human circulatory system. The heart is the power source, the arteries, veins, and capillaries are the wires, and the organs, bones, and muscles are the components such as light sources, resistors, and capacitors. These next questions introduce us to circuit components and wire resistance.

16.5.1.   **Based on the wire resistance equation, what would happen to the resistance in a wire if it doubled in length, would it increase or decrease?**

16.5.2.   **Based on the wire resistance equation, what would happen to the resistance in a wire if it is doubled in thickness, would it increase or decrease?**

16.5.3.   **Based on Table 16.2, which lists resistivity values for certain materials, which two materials would be chosen for making electrical wire? Explain why?**

16.5.4.   **Copper is the standard material used in electoral wires. Discuss three reasons why it is used over other materials.**

The reason circuits are built is to move electricity to do useful things for people. The current flow can do things like light up dark spaces, make noise, make heat, and run

Table 16.2 Electrical Resistivity for common materials

Material	Resistivity at 20 °C (Ω m)
Aluminum	2.8×10^{-8}
Copper	1.7×10^{-8}
Silver	1.6×10^{-8}
Nickel	7×10^{-17}

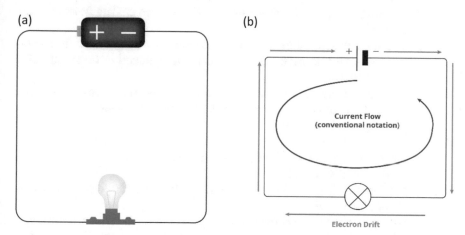

(a)

(b)

Current Flow
(conventional notation)

Electron Drift

Fig. 16.4 (a) Drawing of a simple circuit and (b) circuit diagram with symbols showing two flow notations for circuits

devices. These things slow down or drain a power supply, so they are known as **loads**. To create a circuit, a circular path for the flow of electricity is created by connecting the positive voltage terminal of a power source (for example, a battery) to some conducting item (like a light bulb), and then back to the negative terminal. An example of a simple circuit is shown in Fig. 16.4.

One important thing about Fig. 16.4 is that there are two designations for the electric flow. The red arrow designates *current flow or conventional flow notation* while the green arrows note what is called *electron flow notation*. Conventional flow notation is followed by most electrical engineers and illustrated in most engineering textbooks. However, it does not really matter how the flow of electricity is designated as long as the symbols and model that are used are consistent. The following questions will help us explore the meaning of both notations.

Finally, two common mistakes made with circuits are a short circuit and an open circuit [9]. A **short circuit** is where there is no restriction to current flow, essentially resistance becomes zero. Since there is no resistance the current will be infinite. This is bad because the power source will try to continuously provide current, which could damage electronics, overheat things or drain a battery. Many homes and electronic devices have circuit breakers to prevent damage from short circuits. An **open circuit** is where the loop is not fully connected which leads to the resistance to essentially become infinity. No current flows but a nonzero voltage can appear across parts of the open circuit.

16.5.5. **Looking at Fig. 16.4b, where is the highest amount (surplus) of electric charge in the battery, the positive terminal or negative terminal? (Hint: how many volts does the negative terminal measure)**

16.5.6. **Looking at Fig. 16.4b, where is the lowest amount (deficiency) of electric charge in the battery, the positive terminal or negative terminal? (Hint: how many volts does the negative terminal measure)**

16.5.7. **Looking at Fig. 16.4b, describe how the electric charge flows through a circuit.**

16.5.8. **Based on the previous questions, explain the conventional flow notation in a circuit between the positive and negative terminals of a power source.**

16.5.9. **Looking at Fig. 16.4b, which terminal do you think has the highest amount (surplus) of electrons in the battery; the positive terminal or the negative terminal?**

16.5.10. **Looking at Fig. 16.4b, which terminal do you think has the lowest amount (deficiency) of electrons in the battery; the positive terminal or the negative terminal?**

16.5.11. **Looking at Fig. 16.4b, describe how the individual electrons flow through a circuit.**

16.5.12. **Based on the previous questions, explain the electron flow notation.**

16.5.13. **What is the difference between conventional flow notation and electron flow notation?**

16.5.14. **Based on the definition given above, is there anything resisting the flow of current in a short circuit?**

16.5.15. **If you had a battery and wire, draw a circuit diagram of a short circuit (use a symbol for the diagram).**

In order to analyze simple circuits, there are a few fundamental laws to learn. The first important one is **Kirchhoff's Current Law**. The law was developed by Gustav Kirchhoff, a German physicist. Kirchhoff's Current law is based on the idea that the charge in a circuit is always conserved, even when the wire splits, and the charge travels down different paths [10]. When a wire gets split, the point where the split occurs is called a *junction* or a *node*. In a circuit diagram, a junction is typically represented by a dot or a circle where the lines (wire) change directions. The law states that the current flowing into a junction must equal the current flowing out of it. It could also be stated that the sum of the currents entering and leaving a node equals zero.

The second important fundamental law to learn is **Ohm's Law**. This law was discovered by Georg Simon Ohm, who was previously mentioned. His paper, "The Galvanic Circuit Investigated Mathematically," was published in 1827 and presented the relationship between current (I), voltage (V), and resistance (R) [11]. Ohm's law states that the *current is directly proportional to electrical resistance* *times current*, and thus current is inversely proportional to resistance. The next set of questions helps us to apply Kirchhoff and Ohm's laws to current flow situations.

16.5.16. **Looking at Fig. 16.5a, circle the junction node.**

16.5.17. **Looking at Fig. 16.5a, develop an equation for the currents at the junction using Kirchhoff's Current Law.**

16.5.18. **Based on the previous question, if $I_1 = 1$ A, $I_2 = 5$ A, and $I_3 = 1$ A, determine the missing current value in the circuit in Fig. 16.5a.**

16.5.19. **Based on the explanation above, write out Ohm's law in symbolic form.**

16.5.20. **Based on Ohm's Law, as electric potential increases, what happens to the current, does it increase or decrease?**

16.5.21. **Based on Ohm's Law, as resistance increases, what happens to the current, does it increase or decrease?**

16.5.22. **What are the units of measurement for current, voltage, and resistance? (Recall back to previous sections if necessary)**

16.5.23. **When an electric potential of 1 V exists across an electrical conductor that has a resistance of 1 Ω, how much current will flow through the conductor?**

16.5.24. **The current in a simple circuit is shown in Fig. 16.5b, if the resistance of the lamp is 144 Ω and it is plugged into a wall outlet that supplies 120 V, what is the current that flows through the circuit?**

16.5.25. **The current in a simple circuit is shown in Fig. 16.5b, if the lamp is now plugged into an external battery that supplies 36 V, and the current is 4 A, what is the resistance offered by the lamp?**

16.5.26. **The current in a simple circuit is shown in Fig. 16.5b, if the lamp has a resistance of 8 Ω, and the current that flows is 2 A, what is the amount of voltage needed to be provided by the battery to make the lamp light?**

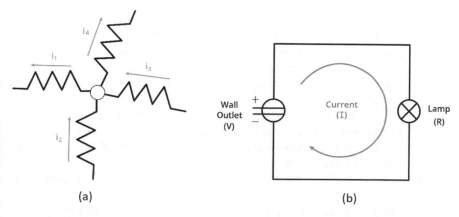

(a)   (b)

Fig. 16.5 (a) Circuit junction shown with current flow directions (b) simple circuit between a duplex outlet and lamp

Recall that in a circuit, wires are connected to a power source and resistive components to control the flow of current. There are two basic arrangements for simple circuits: parallel and series arrangement. A **series** arrangement is when the resistors are connected end to end. In a series circuit, the following occurs:

1. The current that flows through each resistor in the series is the same.
2. There is a voltage drop across each resistor as the current flows through it, and that drop can be determined by Ohm's Law.
3. The total voltage (V_T) supplied by the power source is equal to the sum of the voltage drops across each resistor that is in the series.
4. The total resistance in the circuit (R_T) is equal to the sum of the values of each resistor in the circuit. Thus, the total current in the system can be found by applying Ohm's Law with the total voltage and total resistance.

In a series circuit, if one resistor fails, everything downstream of that element fails because the current gets prevented from flowing to the rest of the elements in the circuit.

A **parallel** arrangement is when multiple resistors are connected between the same two terminals. Thus, the current is divided among the various paths. In a parallel circuit the following occurs:

1. The resistors are connected between the same two points, so the voltage across each of the resistors is the same.
2. The current is divided among the paths, so the sum of the current in each branch is equal to the total current flowing through the circuit. Ohm's law can be used to determine the current flow in each branch.
3. The reciprocal of the total resistance in the circuit ($1/R_T$) is equal to the sum of the reciprocals of the separate resistance values.

In a parallel circuit, if one resistor fails, the entire circuit will not shut down. The current will be diverted to the other working pathways. The next set of questions lets us explore parallel and series circuit arrangements.

16.5.27. **Do you think outdoor lights that are used to decorate a house for the holidays run in a series or parallel arrangement?**

16.5.28. **Do you think the wiring system in a residential house runs in a series or parallel arrangement?**

16.5.29. **For the series circuit shown in Fig. 16.6a, what value is constant V, I, or R?**

16.5.30. **For the series circuit shown in Fig. 16.6a, write an equation to solve for the total resistance of the circuit.**

16.5.31. **For Fig. 16.6a, if the resistors have values $R_1 = 4\ \Omega$ and $R_2 = 12\ \Omega$, what is the total resistance of the circuit?**

16.5.32. **For Fig. 16.6a, if the total voltage supplied is 12 V, calculate the current in the circuit.**

16.5.33. **For Fig. 16.6a, between what terminals (1, 2, or 3) does the voltage change because of Resistor 1?**

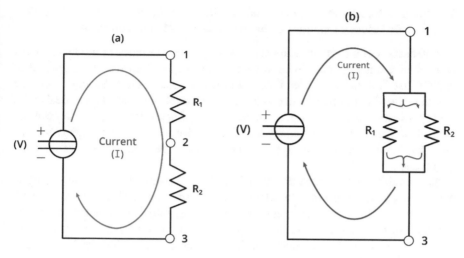

Fig. 16.6 Circuit drawings for (**a**) series and (**b**) parallel arrangements of two resistors

16.5.34. **For Fig. 16.6a, write an equation that would calculate the voltage change (V_1) for the first resistor, based on the value of Resistor 1 (R_1) and the current (I).**

16.5.35. **For Fig. 16.6a, between what terminals (1, 2, or 3) does the voltage change because of Resistor 2?**

16.5.36. **For Fig. 16.6a, write an equation that would calculate the voltage change (V_2) for the second resistor, based on the value of Resistor 2 (R_2) and the current (I).**

16.5.37. **For Fig. 16.6a, if the voltage supplied is 12 V and $R_1 = 4\ \Omega$ and $R_2 = 12\ \Omega$, calculate the voltage drop across each resistor.**

16.5.38. **For Fig. 16.6a, what calculation check can be performed to show that the values calculated in the previous question are correct?**

16.5.39. **For the parallel circuit shown in Fig. 16.6b, what value is constant V, I, or R?**

16.5.40. **For the parallel circuit shown in Fig. 16.6b, write an equation to solve for the total resistance of the circuit.**

16.5.41. **For Fig. 16.6b, if the resistors have values $R_1 = 4\ \Omega$ and $R_2 = 12\ \Omega$, what is the total resistance of the circuit?**

16.5.42. **For Fig. 16.6b, if the circuit is connected to a 60 V battery, what is the total current in the line?**

16.5.43. **For Fig. 16.6b, write an equation that would calculate the current flowing (I_1) through the first resistor line.**

16.5.44. **For Fig. 16.6b, write an equation that would calculate the current flowing (I_2) through the second resistor line.**

16.5.45. **For Fig. 16.6b, if the circuit is connected to a 60 V battery and resistors values $R_1 = 4\ \Omega$ and $R_2 = 12\ \Omega$ calculate the current flowing through each line?**

16.5.46. **For Fig. 16.6b, what calculation check can be performed to show that the values calculated in the previous question are correct?**

Another aspect of a circuit to understand is the power consumed or produced by the components. It is important to know because **power** is a measure of the amount of energy that is transferred over time, and energy costs money. So power can be directly related to the cost of an electronic component or circuit. Power deals with energy, which will be discussed in detail in next chapter, but energy can never be created nor destroyed. It only changes to different forms. In a circuit, energy is often changed from electrical energy to mechanical, thermal, or electromagnetic (light) energy.

In a circuit, a component that transforms some energy into electric energy is called a **producer**. For example, a battery transforms chemical energy into electrical energy. A component that changes electric energy into a different type of energy is a **consumer** [12]. There are many consumers in a circuit. For example, a light bulb is a resistive element that transforms electric energy into both light and heat energy.

Recall the SI unit for energy is Joule (J). Since power measures how much electric energy is transferred, and how fast the transfer happens, the unit for electric power is the Joule per second (J/s), also called a **Watt**. It is named after Scottish mechanical engineer, chemist, and inventor James Watt [13]. Since electric energy consumption occurs over a wide range of applications, standard prefixes are typically placed before the W in order to quantify the magnitude of usage. For example, microcontrollers such as Arduinos operate in the μW or mW range, whereas the energy consumption of a building is typically in the kW range, and the operation of a professional sports stadium light would operate in the MW or GW range. This next inquiry will help us calculate power in terms of the basic terms applied in a circuit, namely current, voltage, and resistance.

16.5.47. **For the examples of prefixes given for power operation (μW, mW, kW, MW, and GW), how many watts is each of them? (Hint: Look back at Chap. 10 if needed)**

16.5.48. **What is the fundamental SI unit of the unit Watt?**

16.5.49. **What is the fundamental SI units of the unit Volt (hint: Check back to question 16.2.4)?**

16.5.50. **What is the fundamental SI unit of the unit Ampere (hint: Check back to questions 16.2.3)?**

16.5.51. **Based on the previous three questions, develop a relationship between the units of power, voltage, and current.**

16.5.52. **Based on the previous question, develop a relationship between power (P), current (I), and voltage (V).**

16.5.53. **Based on the previous question, use Ohm's law to develop a relationship between power (P), voltage (V), and resistance (R).**

16.5.54. **Based on question 16.5.52, use Ohm's law to develop a relationship between power (P), Current (I), and resistance (R).**

16.5.55. **If a simple circuit is powered by a 9 V battery and is connected to an 11 Ω resistor, calculate the power across the resistor.**

16.5.56. **What form of energy is that resistor transforming the electric energy into?**

16.5.57. **Resistors have *power ratings* on them to alert users of unwanted energy transfer. If a resistor has a power rating of 0.5 W, would the resistor be able to be used in the simple circuit in question 16.5.55? Explain why or why not.**

All the concepts previously discussed; current, voltage, resistance, Kirchhoff's law, Ohm's law, series and parallel arrangements, and power relationships can be used to analyze simple circuits. The next set of questions gives us practice doing just that.

16.5.58. **For the circuit drawing in Fig. 16.7 find the total resistance of the circuit.**

16.5.59. **For the circuit drawing in Fig. 16.7 find the total current drawn by the circuit.**

16.5.60. **For the circuit drawing in Fig. 16.7 find the voltage drop across the first resistor.**

16.5.61. **For the circuit drawing in Fig. 16.7 find the current drop across R_2 and R_3.**

16.5.62. **For the circuit drawing in Fig. 16.7 what is the power consumed by the circuit?**

16.5.63. **A 9 V battery supplies 19,440 J of energy. How long will the circuit run before the battery is drained? (Recall general relationship between power and energy).**

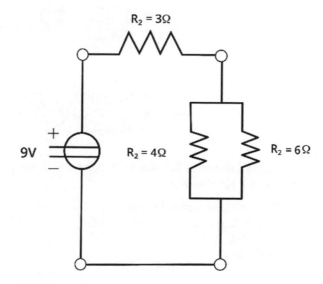

Fig. 16.7 Circuit drawing of simple circuit configuration with a power supply and resistors

End of Chapter Questions

IBL Problems

IBL 1: In the United Kingdom, the equation for mains electrical supply is $V(t) = 325\sin(2\pi 50t)$. Use this to answer the following questions

1. What is the frequency of the current in typical UK houses?
2. Explain what the value 325 stands for.
3. Based on the previous questions, what is the equivalent RMS voltage used in the United Kingdom?
4. What is the equivalent DC voltage?

Practice Problems

1. For the following pairs of appliances supplied with primary voltage in a domestic residence in the United States, how large of a breaker is needed to that both appliances can be plugged in simultaneously? Would the pairing trip a 15 amp GFI?

   • Coffee Maker (200 W) and Microwave Oven (1500 W)
   • Toaster (1200 W) and Microwave Oven (1500 W)
   • Hair dryer (2000 W) and Electric Curlers (300 W)
   • Treadmill (2000 W) and Dehumidifier (900 W)

2. For the following resistors, calculate the resistance and tolerance values.

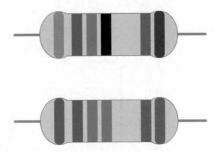

3. Calculate the current that flows through the following light bulbs if each is connected to a 120V source.

   • 75W incandescent light bulb
   • 53W halogen incandescent light bulb
   • 19W CFL
   • 17W LED

4. For the circuit shown calculate (1) the total resistance, (2) current flow for the circuit, and (3) current flow for each circuit branch.

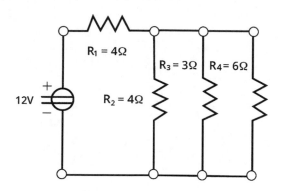

References

1. Huray, Paul G., 1941- (2010). Maxwell's equations. Hoboken, N.J.: Wiley. pp. 8, 57. ISBN 978-0-470-54991-9. OCLC 739118459.
2. Walker, Jearl; Halliday, David; Resnick, Robert (2014). Fundamentals of physics (10th ed.). Hoboken, NJ: Wiley. p. 614. ISBN 9781118230732. OCLC 950235056.
3. Horowitz, Paul; Hill, Winfield (2015). The art of electronics (3rd ed.). Cambridge University Press. ISBN 978-0-521-80926-9.
4. Pancaldi, Giuliano (2003). Volta, Science and Culture in the Age of Enlightenment. Princeton Univ. Press. ISBN 978-0-691-12226-7.
5. Skrabec, Quentin R. (2012). The 100 Most Significant Events in American Business: An Encyclopedia. Santa Barbara, California: ABC-CLIO. ISBN 978-0-31339-863-6.
6. "Ohm, Georg Simon". Encyclopædia Britannica. 20 (11th ed.). Cambridge University Press. p. 34. https://en.wikisource.org/wiki/1911_Encyclop%C3%A6dia_Britannica/Ohm,_Georg_Simon viewed on 2/15/2021.
7. Chisholm, Hugh, ed. (1911). "Faraday, Michael". Encyclopædia Britannica. 10 (11th ed.). Cambridge University Press. pp. 173–175. the 1911 Encyclopædia Britannica. https://en.wikisource.org/wiki/1911_Encyclop%C3%A6dia_Britannica/Faraday,_Michael viewed on 2/15/2021.
8. Don Johnson, Rice University (2014). Fundamentals of Electrical Engineering I. OpenStax CNX. ISBN 13: 9781300160137.
9. Saeed Moaveni (2005). Engineering Fundamentals: An Introduction to Engineering (2nd ed.). Thompson. ISBN: 0-534-42459-7.
10. Tony R. Kuphaldt (2012). Lessons in Electric Circuits Volume I - DC. Koros Press. ISBN-13: 978-1907653087.
11. Ohm, G. S. (1891). The galvanic circuit investigated mathematically. New York: D. Van Nostrand.
12. Electric Power. https://learn.sparkfun.com/tutorials/electric-power. viewed March 3, 2021.
13. J.Richard Elliott and Carl T. Lira (2012), Introductory Chemical Engineering Thermodynamics, 2nd edition. Pearson. ISBN-13: 978-0-13-606854-9

Chapter 17
Energy, Power, and Efficiency

Abstract Energy is a fundamental concept that emerges in many engineering analyses. In many engineering classes, particularly in thermodynamics and solid mechanics, energy will need to be calculated. There are many forms of energy. We briefly touched upon some in previous chapters. Specifically, thermal energy when we discussed temperature and electrical when we discussed circuits. In this chapter, we will work to define energy and work to calculate energy in all its many usable forms. Then we will discuss the relationships between power, work, and energy. Finally, we will define the term efficiency and explore how it relates to different systems including heating and cooling.
By the end of this chapter, students will learn to:

- Define what energy is and its role in society.
- List the many forms of energy and energy sources.
- Describe mechanical energy.
- Calculate mechanical energy in its different forms.
- Explain the relationship between work and energy.
- Calculate the work done on an object or the distance traveled by an object when work is done on it.
- Define Chemical Energy and explain examples of chemical energy.
- Explain how the units of chemical energy and thermal energy are related.
- Establish an understanding of the laws of Conservation of Energy as they relate to energy conversion and transfer.
- Define power, explain its common units, and understand how it relates to work and energy.
- Calculate the power required for a task or engineering application.
- Explain the limitations of energy conversion in terms of efficiency.
- Calculate efficiency or power of a system or machine.

© Springer Nature Switzerland AG 2022
M. Blum, *An Inquiry-Based Introduction to Engineering*,
https://doi.org/10.1007/978-3-030-91471-4_17

17.1   Importance of Energy

Energy is an intangible concept because you cannot physically touch it. The results of energy can be seen when it is produced. Energy is defined as a property that is transferred to an object in order to perform work on (or to heat) another object [1]. Simply put, energy is the ability of *something* to *move* (*or heat up*) something else. Engineers are concerned with how energy is stored, produced, and transferred. So, first, how is energy produced? There are many ways energy can be produced, but the source of energy for most life on earth is the sun. Second, how is energy transferred? There are many forms of energy and energy can be converted from one form to another through various physical processes. These types of energies come with their own terms and definitions, depending on the form of energy being discussed. Finally, how is energy stored? Energy is stored in various forms to be discussed. For engineers, the ultimate goal is to efficiently convert the stored energy to a desirable form so it can be used. These next few questions will help us begin to explore the concept of energy.

17.1.1. **Explain in your own words why energy is important in society.**
17.1.2. **List as many forms of energy as you can think of (you may use the ones previously discussed) (hint: there are eight major ones).**
17.1.3. **List as many as you can think of: examples of energy sources currently available in society.**

17.2   Category of Energy I: Mechanical Energy

To help make quantifying energy easier, its different forms are classified into different categories. One of the first major categories is mechanical energy. **Mechanical energy** is the ability of an object to do work. Recall Chap. 14 when we defined **work (W)**. Work is when a force moves an object over some distance. The object gains energy because work is done upon it. Mechanical energy can be possessed by an object due to either (1) its motion or (2) due to its position. Mechanical energy has two major forms—**kinetic energy** and **potential energy** [2].

Kinetic energy (*KE*) is the form of energy an object has when it is moving. When work is done on an object, it changes the object's kinetic energy. There are two main forms of kinetic energy;

1. **Translational (KE$_T$):** this is the energy due to the linear motion of an object from one place to another. Mathematically, the change in translational KE$_T$ of an object with mass (m) sliding from one position with an initial velocity (v_i) to a second position moving with a final velocity (v_f) can be written as $\Delta KE_T = \frac{1}{2}m\left(v_f^2 - v_i^2\right)$. When work is done on an object it changes the kinetic energy of the object, so we can also say that from an initial to a final position is equal to the change in translational kinetic energy, namely $W_{i \to f} = \Delta KE_T$.

2. **Rotational (KE$_R$)**: this is the energy an object has due to its spinning about an axis. Rotational KE$_R$ of an object depends on its shape, represented by the area moment of inertia (I) and its angular velocity (ω). Mathematically, the change in rotational KE of an object that is first rotating with initial angular velocity (ω_i) and then a final angular velocity (ω_f) can be written as $\Delta KE_R = \frac{1}{2}I\left(\omega_f^2 - \omega_i^2\right)$. Sometimes it is also called angular kinetic energy.

Potential Energy (PE) is energy stored in an object as the result of a change of position. Work is done to deviate the location of the object with respect to some origin. There are two main forms of potential energy;

1. **Gravitational (PE$_G$)**: this is the energy stored in an object as the result of a change in its vertical position or height because mechanical work is performed to overcome the gravitational force of the earth that wants to pull the object back down. PE$_G$ of an object depends on the mass of the object (m), the change in the distance it has been lifted (Δh), and earth's gravity (g). Mathematically the equation is written as $\Delta PE_G = mg(\Delta h)$. To determine the gravitational potential energy of an object, an origin (zero height position) must be assigned. Assigning an origin is arbitrary and typically the ground is labeled as the origin.
2. **Elastic (PE$_E$)**: this is the energy stored in objects that are elastic (meaning they can stretch and compress and then return to their original shape). The amount of PE$_E$ stored is related to the amount of stretch. Mathematically, elastic potential energy is modeled from spring force, which follows hooks law (recall back our discussion on spring force in Chap. 14). For a spring that is stretched from an initial position (x_i) to a final position (x_f) with a spring constant (k), the equation is expressed as $\Delta PE_E = \frac{1}{2}k\left(x_f^2 - x_i^2\right)$. Please note that just like PE$_G$, an origin must be set to track the deviation of the spring position. In most cases, the origin is the spring equilibrium position. This is the unstretched or compressed position that the spring naturally assumes when no force is applied to it. Sometimes the initial position (x_i) is the origin, sometimes it is not. In order to accurately calculate PE$_E$, the spring's displacement needs to be correctly measured.

The next set of questions helps us explore the concept of mechanical energy and calculate energy of its different forms.

17.2.1. **Describe the form of mechanical energy (KE$_T$, KE$_R$, PE$_G$, PE$_E$) each of these objects have: a speeding car, a drawn archery bow, a skydiver about to jump out of a plane, a spinning ice skater.**
17.2.2. **Write the mathematical formula for work (W) in terms of force (F) and distance (d).**
17.2.3. **If you push a heavy box across the floor, which will make you more tired; pushing it 5 ft or pushing it 25 ft? Explain your answer.**

For the next six questions, use the following scenario shown in Fig. 17.1a: A student on a bicycle is traveling at a speed of 12 mph. Together the student and bicycle weigh 148 lb. The cyclist needs to stop at a stop sign that is coming up in 300 ft.

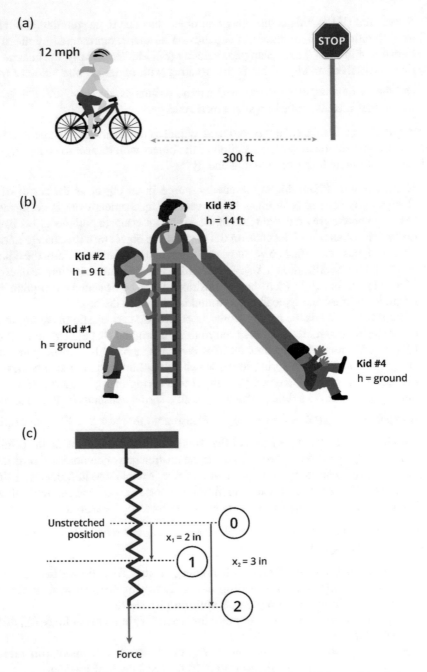

Fig. 17.1 (**a**) Student on bicycle needs to stop at the stop sign, (**b**) Children waiting their turn to go down a slide, and (**c**) spring being stretched past its equilibrium point

17.2.4. **What is the initial velocity of the scenario in standard US units?**

17.2.5. **What is the final velocity of the scenario in standard US units?**

17.2.6. **In the scenario, what is the mass of the student and the bicycle? (Recall how to convert weight to mass for US units).**

17.2.7. **In the scenario, what is the change in total translational kinetic energy so the bike comes to a stop?**

17.2.8. **In the scenario, what is the force needed to bring the bike to a stop at the stop sign?**

17.2.9. **For the scenario, what happens to the initial kinetic energy? What other form of energy is it converted to?**

17.2.10. **Which will have a higher PE_G, a heavy object or a light object?**

17.2.11. **Which will have a higher PE_G, a book on the bottom shelf of a bookcase or a book on the top shelf?**

17.2.12. **For the picture given in Fig. 17.1b, determine the total potential energy for each student on the slide (Assume the children all have the same weight of 50 lb.)**

17.2.13. **For Child #4 what happens to the initial potential energy? What other form of energy is it converted to?**

17.2.14. **For the spring shown in Fig. 17.1c, determine the change in elastic energy of the spring when it is stretched from position 0 to position 1 (spring constant is 625 lb$_f$/in.).**

17.2.15. **For the spring shown in Fig. 17.1c, determine the change in elastic energy of the spring when it is stretched from position 1 to position 2 (spring constant is 625 lb$_f$/in.).**

17.2.16. **For the spring shown in Fig. 17.1c, determine the change in elastic energy of the spring when it is stretched from position 0 to position 2 (spring constant is 625 lb$_f$/in.).**

17.2.17. **What is the relationship between the answers calculated in questions 17.2.14, 17.2.15, and 17.2.16? What does this tell us about the spring?**

17.2.18. **For the spring shown in Fig. 17.1c, what happens to the total elastic energy from position 0 to position 1 if position 1 compresses the spring 2 in.? (Spring constant is 625 lb$_f$/in.)**

17.3 Conservation of Mechanical Energy

Since we know that energy can only change form, if we neglect other type of energy, we can develop a statement for the conservation of mechanical energy. The **conservation of mechanical energy** states that the total mechanical energy of a system is constant as long as there is no thermal energy (heat transfer). It can also be stated that the sum of the kinetic and potential energy of an object is zero. The next set of questions lets us develop and study the conservation of mechanical energy.

17.3.1. **Based on the previous section and the above paragraph, write a mathematical expression for the definition of the conservation of mechanical energy**

For the next three questions, use the following scenario shown in Fig. 17.2: A 2 kg cart is pushed up against a spring and it deflects 0.1 m in position 1. Then the cart is released from rest and begins to only slide. The spring constant is 2000 N/m and assume energy is conserved and friction is neglected.

17.3.2. **For the scenario shown in Fig. 17.2, what types of mechanical energies are present?**

17.3.3. **Based on the previous answer, write a mathematical expression for the conservation of mechanical energy in the given scenario.**

17.3.4. **Based on the previous answer, calculate how high the cart travels when it reaches a velocity of 2.5 m/s.**

17.4   Work

Recall back to Chap. 14 when mechanical work (*W*) was discussed. Work was performed when an object is moved a certain distance by a force. When the force moves the object, energy is transferred to it. Work and energy are directly related to each other because work is essentially energy in motion. We have also previously discussed the different forms energy in Sect. 17.2. Recall that we described energy in motion as kinetic energy. Thus, when work is done on an object its kinetic energy changes. *The net work done by a force on an object is equal to the change in its translational kinetic energy.* This section lets us explore the relationship between work and energy.

17.4.1. **Write the mathematical equation for work (Hint: look back at Chap. 14 for help if needed).**

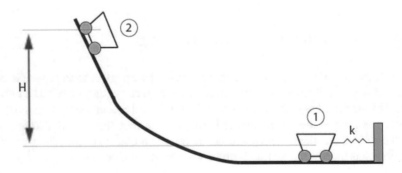

Fig. 17.2 Illustration of cart attached to spring

17.4.2. **Write the mathematical equation for kinetic energy of an object (Hint: look back at Sect. 17.2 for help if needed).**

17.4.3. **The sentence in *Italics* above explains the relationship between work and kinetic energy, write a mathematical expression based on that sentence.**

17.4.4. **You are moving a couch and you can push with a force of 10 lb. One possible position in the room requires you to slide the couch 6 ft, a second possible position requires you to slide the couch 20 ft. How much work is required for you to move the couch to position 1? How much work is required for you to move the couch to Position 2?**

17.4.5. **For the previous question, moving the couch to which position will make you more tired; position 1 or 2? Explain why.**

17.4.6. **A bicycle is traveling at a velocity of 20 mi/h. The mass of the bike and rider is 200 lb. What is the force needed to bring the bike to a full stop at a distance of 50 ft? (hint: Be aware of units, particularly for mass!)**

17.4.7. **From the previous question, what happens to the initial energy the bike and rider had? What changes in the form of energy occur?**

17.5 Category of Energy II: Thermal Energy

Recall from when we discussed temperature parameters in Chap. 15 we spoke about how heat transfer was the transfer of thermal energy. Thermal energy is the energy contained within an object due to its molecular activity. Thermal energy is responsible for that object's temperature. Having objects with different temperatures results in heat transfer. In Chap. 15 we discussed (1) some rules of thermal energy transfer, (2) different modes of thermal energy transfer, and (3) the units of thermal energy.

Thermal energy plays a major role in mechanics problems because it is the foremost nonconservative energy outlet. Based on the conservation of energy, energy is always conserved. However, in many mechanics problems energy never fully transfers from potential to kinetic and vice versa. In real-work physical systems, almost every transfer of energy results in some thermal energy. Many times, the amount is very weak and remains close to the energy of the surrounding environment, but it does occur. The two major culprits of this thermal energy creation are (1) dry *friction* and (2) *drag*.

The nonconservative force from friction was introduced in Chap. 14. **Dry friction** force occurs because a force resists the relative motion between two solid surfaces due to the interaction of irregular features of the surfaces of the bodies. There are two mechanisms of dry friction, static and kinetic. **Static** friction occurs when two objects are in contact but do not move relative to each other. For example, the static friction force prevents people from slipping when they walk. **Kinetic** friction (also known as dynamic or sliding friction) occurs when two objects move (slide) relative to each other and rub together. The static force of friction does not generate thermal energy. Kinetic friction results in heat generation.

Dry friction was modeled by Charles-Augustin de Coulomb, a French military engineer and physicist [3]. Coulomb's law of friction states that for static friction $F_s = \mu_s N$ and for moving objects $F_f < \mu_k N$. Where F_f is the friction force exerted by each surface on the other. The friction force is always parallel to the surface and in the opposite direction of the applied force. The coefficient of friction is denoted by the parameter μ. This is an empirical property based on the contacting materials. Static friction coefficient is μ_s and the kinetic friction coefficient is μ_k. Finally, N denotes the normal force exerted by each surface on the other. The normal force is always directed perpendicular to the contacting surfaces. Engineers draw Free Body diagrams to identify the normal and frictional forces for friction calculations.

Another nonconservative force is drag. **Drag** is an aerodynamic force that is caused by the motion of a body through a fluid. Like dry friction, drag is a resistive force that acts in the opposite direction to the fluid velocity. A standard *drag equation* is used to calculate the resistance experienced by an object moving through a fluid. The drag equation was developed by British Scientist Lord John William Sturt, third Barron of Rayleigh [4]. The equation states that the drag force (D), exerted on a body traveling through a fluid can be found from $D = \dfrac{1}{2} C \rho A v^2$. Where C is known as the drag coefficient. This is a dimensionless number that is dependent on the shape of the body. Typical values range from 0.4 to 1.0 for different fluids (such as air and water) [3]. The density of the fluid through which the body is moving is denoted by ρ, and the velocity of the body relative to the fluid is denoted as v. The projected cross-sectional area of the body (A) is perpendicular to the flow direction, namely the velocity.

Recall energy is the capacity to do work, so both friction and drag can be related to work needed to be done to overcome the forces or calculate the energy generated from the work done by the forces. The next set of questions lets us explore thermal energy, friction, and drag.

17.5.1. **Recall from Chap. 15, how does heat flow from one object to another based on the temperatures of the objects?**

17.5.2. **Recall from Chap. 15, what were the three main modes of thermal energy (heat) transfer?**

17.5.3. **Recall from Chap. 15, what were the SI and US customary units of thermal energy?**

17.5.4. **A person is sliding a large box across a flat surface as shown in Fig. 17.3. Draw a free body diagram of the system (Recall steps from Chap. 14).**

17.5.5. **Based on the Free Body diagram that was drawn, circle the friction for and the normal force.**

17.5.6. **Based on the Free Body diagram that was drawn, calculate the normal force.**

Fig. 17.3 Women pulling a box opposed by dry friction

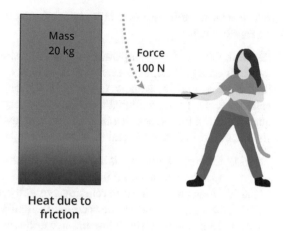

Mass
20 kg

Force
100 N

Heat due to
friction

17.5.7. **Based on the Free Body diagram that was drawn, if the kinetic friction coefficient is 0.3, calculate the friction force.**

17.5.8. **The change in total thermal energy (ΔE_T) equals the total work done by friction. Based on the definition of work, develop an equation for ΔE_T.**

17.5.9. **If the person in Fig. 17.3 pulls the box with a constant velocity at a distance of 75 m, how much thermal energy will be transferred between the box and the floor?**

17.5.10. **A person is swimming laps in a pool (water density is 1000 kg/m³ and drag coefficient is 0.45). We can assume the frontal area of the swimmer exposed to the water is 8 cm². What is the drag force on the swimmer if they are swimming at a constant speed of 1.5 m/s?**

17.5.11. **Based on the previous question, how much thermal energy is required for the swimmer to complete a 1.0 km swim if they keep the same pace?**

17.5.12. **What will happen to the energy the swimmer supplies?**

17.6   Category of Energy III: Chemical Energy

The final category of energy to address is chemical energy. Recall back to Chap. 13 when we discussed mass that everything is made up of atoms and molecules. When atoms are connected there is energy stored in the bonds. This is known as **Chemical Energy**. When the bonds break, energy is released, and to form chemical bonds energy needs to be absorbed. The creation or breaking of bonds is known as a **chemical reaction**. In a chemical reaction process atoms become rearranged or changed, resulting in the production of a new substance. There are two types of chemical reactions:

1. An **Endothermic** reaction is a chemical reaction that absorbs energy from its environment.

2. An **Exothermic** reaction is a chemical reaction that releases energy into its environment.

Recall in Chap. 15 on Temperature we learned about the different units of thermal (or heat) energy. Chemical energy is measured in the same units that nutrition or food is measured in, namely **Calories (Cal)**, but thermal energy is measured in **calories (cal)**. Then recall that the SI unit of energy is the Joule (J). The following questions help us discuss chemical energy and investigate the relationship between chemical, thermal, and general SI units of energy.

17.6.1. **Chemical energy is stored in atomic and molecular bonds, it is a kinetic or potential form of energy?**

17.6.2. **Does an exothermic reaction typically feel hot or cold?**

17.6.3. **Does an endothermic reaction typically feel hot or cold?**

17.6.4. **List whether the following are endothermic or exothermic reactions: (1) wood in a fireplace, (2) eating food, (3) melting ice cubes, and (4) battery in a calculator.**

17.6.5. **For the reactions in the previous question, explain what changes occur in the forms of energy.**

17.6.6. **How many thermal energy units (cal) are in 1 chemical energy unit (Cal)? (hint: look back to Chap. 15 if needed)**

17.6.7. **Which energy unit is the larger amount; thermal energy unit or chemical energy unit?**

17.6.8. **How many Joules is equal to 1 cal of thermal energy? (hint: look back to Chap. 15 if needed).**

17.6.9. **How many Joules is equal to 1 Cal is Chemical energy?**

17.6.10. **Name any other forms of chemical energy you can think of.**

17.7   First and Second Law of Thermodynamics

Recall we have established that energy can never be created nor destroyed. It can only be transferred to another form. Also, in Sect. 17.3 we discussed the conservation of mechanical energy which assumed there was no heat energy or work involved with the system. Now work and heat will be introduced.

Before developing a mathematical statement, we need to define what a **system** is. Recall back to Chap. 13, when we were discussing mass and we developed a definition for a control volume. A system is like control volume, in that it is a defined space of special interest for the analysis. It typically has a clear boundary that can exchange energy and sometimes material (mass/matter). There are three types of thermodynamics systems based on this exchange:

- Open System—A open system can freely exchange both energy and mass over its borders with its surroundings.
- Closed System—A closed system exchanges energy over its borders but not mass with its surroundings.

- Isolated System—An isolated system does not exchange energy or matter over its borders with its surroundings. A perfectly isolated system does not really exist. It is typically an assumption for mathematic calculations.

Closed systems will be the focus of our analysis for this section due to the complexity of including mass transfer analysis for open systems. For a closed system, the **first law of thermodynamics** states that the internal energy of a system is equal to the work that is being done on the system, plus or minus the heat that flows in or out of the system. This is expressed mathematically as $\Delta E = Q_T - W_T$.

Where ΔE is the total change in energy of the system. It is the sum of all forms of energy in the system namely elastic, potential, kinetic, chemical, etc. For mechanics problems, it is assumed that no nuclear, chemical, or electrical energy is involved so the total energy is typically represented by the sum of kinetic, potential, and **internal energy of the system (U)**. It is written as $\Delta E = \Delta KE + \Delta PE + \Delta U$. When the closed system does not move, the kinetic and potential energy is zero then $\Delta E = \Delta U = U_{final} - U_{inital}$.

The term Q_T represents the net energy transferred into the system $Q_T = \sum Q_{in} - \sum Q_{out}$. The sign convention is very important here. Any heat transferring *into* the system is considered *positive* and any heat *leaving* the system is considered *negative*. The reason for this convention is to show that heat transferring into the system increases the total energy of the system and decreases the total energy when heat transfers out.

The term W_T represents the net work done by the system $W_T = \sum W_{out} - \sum W_{in}$. Again, sign convention is important. Any work being *on* the system is *negative* and any work being *done by* the system is *positive*. The reason for this convention is to show that work done on the system increases the total system energy and work done by the system decreases the total energy.

The next set of questions explore the first law of thermodynamics.

17.7.1. **Label the following examples shown in Fig. 17.4a as either an open, closed, or isolated system: a pot of boiling soup with no lid; a pot of boiling soup with a tight lid on; soup in a perfectly insulated thermos.**

17.7.2. **The piston-cylinder system shown in Fig. 17.4b, what would be defined as the system?**

17.7.3. **Looking at the piston-cylinder system shown in Fig. 17.4b, if the piston does not move and the initial internal energy of the system was 110 J and the final internal energy of the system was 80 J, write an equation to solve for the heat transfer of the system.**

17.7.4. **Looking at the piston-cylinder system shown in Fig. 17.4b, if the piston does not move and the initial internal energy of the system was 110 J and the final internal energy of the system was 80 J, was heat added to the system or removed from the system? How much was added or removed?**

17.7.5. **Looking at the piston-cylinder system shown in Fig. 17.4b, if the piston is moved to the left at a distance 0.6 m with an external force of 30 N, is the work term negative or positive?**

(a)

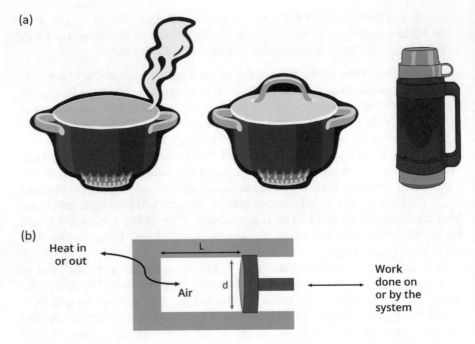

(b)

Fig. 17.4 (a) Example of open, closed, and isolated systems. (b) Picture of piston-cylinder

17.7.6. **Looking at the piston-cylinder system shown in Fig. 17.4b, if the piston is moved to the left at a distance 0.6 m with a force of 30 N, and this is an isothermal process (meaning there is no change in internal energy), how much heat is removed from the system?**

The second law of thermodynamics is more of a verified hypothesis than a law. The first law of thermodynamics deals with the quantity of energy in a system, the second law is about the quality of energy in that system. In 1865 the Germain physicist, Rudolf Clausius in connection with Lord Kelvin stated that heat does not spontaneously flow from a colder body to a hotter body [5]. This statement became the basis for the Second Law. The **second law of thermodynamics** states that in an isolated system, any natural process in that system progresses in the direction of increasing disorder. That disorder, randomness, and uncertainty in a system are defined as **entropy**. It is considered a property of a system and is mathematically denoted with a **S**. This lets the second law be written as $\Delta S = \Delta Q/T$. So, the change in entropy is equal to the change in heat transfer divided by the temperature. For physical processes, there are two outcomes:

- Reversible Process—A reversible process is where the initial and final states of the system are equal ($S_{inital} = S_{final}$). Then the combined entropy of the system and the environment remains constant.
- Irreversible process—An irreversible process is where the final entropy is greater than the initial entropy of the system ($S_{final} > S_{initial}$). Then combined entropy of the system and the environment increases.

Computations involving the second law of thermodynamics will be taught in an upper-level thermodynamics class, but the next question investigates the concept.

17.7.7. **The following processes and interactions between the system and environment, based on the second law of thermodynamics, state whether the process is reversible or irreversible.**

- **A cup of hot tea left on the counter to cool on its own**
- **Extension of spring**
- **Frictionless motion of a solid**
- **Relative motion of a solid with friction**
- **Gas diffusing through a membrane**

17.8   Power

Why is it that a person who runs up a hill is sweaty and tired as opposed to a person who walks up the same hill? If they both weigh the same amount and they both traveled up the same distance, they performed the same amount of work. So why is the person who ran all sweaty and tired? The answer lies in the definition of the term **Power**. Power is defined as the rate of doing work, or the energy spent per time. Many devices, such as cars, refrigerators, furnaces, and lightbulbs, are given power ratings. This indicates the rate at which the machine can do work on other objects. The units of power were established by the Scottish mathematician and engineer James Watt [6]. He created a definition of power based on how a steam engine could replace the work done by a horse. Hence, the unit **horsepower (hp)** was created. Later in 1908, the unit **watt (W)** was accepted as the standard SI unit of power. Both watt and horsepower are still prevalently used today, with Watt being the SI unit for power and horsepower being the US unit. One horsepower is equivalent to 735.5 W. Another unit, **Btu per hour (Btu/h)**, is used in heating, ventilating, and air-conditioning (HVAC) applications to represent the heat loss or gain from a building. The relationship between watts and Btu/h was discussed in Chap. 16 on Temperature. This next set of questions helps us define power, explore the relationship between units, and calculate the power required for certain tasks.

17.8.1. **Write a mathematical expression for power in terms of work and time.**
17.8.2. **Recall and write the definition of work.**
17.8.3. **Based on the definition of power and work, what are the fundamental dimensions of power (Recall Chap. 11 and the seven dimensions).**
17.8.4. **What are the fundamental US and SI units for power?**
17.8.5. **How much power is required to lift a box that weighs 110 lb 5 ft in the air in 1 s. (This is the definition of horsepower.)**
17.8.6. **How much power is required by an elevator to lift ten people, with an average weight of 135 lb per person a vertical distance of 16 ft in 2 s? How much horsepower is that?**

17.8.7. **A student in gym class is performing a chin-up. Her body mass is 40 kg and she lifts herself a distance of 0.3 m in 2 s. What is the power being delivered by the student's muscles?**

17.8.8. **Recall, what is the definition of velocity?**

17.8.9. **Using the previous question, is there another way to write the power equation?**

17.9   Efficiency

Under ideal conditions, all the energy that was put into a project or machine would be fully converted and used to accomplish the desired task. However, this can never happen. As we have spoken about earlier in this chapter, during every form of energy conversion, there are always losses that occur. To quantify how well a machine or system is operating, we calculate its *efficiency*. **Efficiency**, denoted by the Greek letter eta η, measures how much energy or power is lost during a process, and how much energy or power is being effectively utilized for the desired outcome. The efficiency of different systems or devices is calculated in different ways depending on the application, but all the equations deal with the terms input, output, and loss.

- **Input** is the quantity of whatever (energy, power, heat, work, etc.) required by the mechanism to accomplish a task or operate the device.
- **Output** is the quantity of whatever (energy, power, heat, work, etc.) that is actually applied to accomplish a task or operated the device.
- **Loss** is the quantity of whatever (energy, power, heat, work, etc.) wasted during the operation.

Typically, efficiency equations are the ratio of the output over the input. When devices and machines are given power ratings, the power is referred to as the input power, not the output power.

17.9.1. **Can we ever have a system or machine that operates at 100% efficiency: yes or no?**

17.9.2. **Based on the previous question, come up with a range of values for efficiency; in terms of percent and in terms of non-integer numbers.**

17.9.3. **If a machine operates at 80% efficiency, how much energy is lost during the operation?**

17.9.4. **Write a mathematical expression for power efficiency (η).**

17.9.5. **An incandescent light bulb works by heating a piece of wire inside the bulb to such a high temperature that it glows. It has an efficiency of 5%. Explain in terms of percentages, input/output and loss, where 100% of the input power goes.**

17.9.6. **Based on the previous question, what is the relationship between input power, output power, and loss?**

17.9.7. **A 50 W incandescent light bulb is turned on for 20 min. How much heat energy (in Joules) is lost during that period?**

17.9.8. **Explain the relationship between the input and output power for a system if it were 100% efficient**

End of Chapter Problems

IBL Problems

IBL1: Recall that work is equal to the change in translational kinetic energy. If a constant force (F) is applied to a body at rest such that the body moves a distance (d) with a constant acceleration (a) and a linearly increasing velocity (v), answer the following questions:

1. Write an equation for the average velocity (v) of the body in the scenario.
2. Using dimensional analysis, develop an equation to calculate the distance (d) the object traveled based on the average velocity (v) of the object and the time (t) the object traveled.
3. Write an equation that related the force (F) to the constant acceleration of the body (a).
4. What is the definition (equation) for work?
5. Develop an equation for work in terms of mass, acceleration, velocity, and time.
6. What is acceleration? (Answer: change in velocity over change in time)
7. Write the definition of acceleration in terms of the velocities of the body
8. Using the previous answers develop a relationship for work in terms of mass and velocity.
9. Write an equation for the translational kinetic energy of the body.

IBL2: A 15 lb bowling ball with 8.5 in. diameter is thrown down the lane where it hits the pins going an average speed of 17 mph with a rotational speed of 250 rpm. Answer the following questions:

1. When the bowling ball is traveling down the lane, is it translating, rotating, or both?
2. What is the mass of the bowling ball?
3. What is the linear velocity of the ball in ft/s?
4. What is the angular velocity of the ball in rad/s?
5. What is the equation for the area moment of inertia for a sphere?
6. Calculate the moment of inertia for the bowling ball.
7. When the ball is traveling down the lane what type of energies does it have?
8. Using the previous question, write an equation for the total energy of the bowling ball as it travels down the lane.
9. Calculate the total energy of the bowling ball.

Practice Problems

1. As shown in Fig. 17.2, a 2 kg cart is pushed up against a spring and it deflects 0.1 m in position 1. Then the cart is released from rest as shown in the figure. The spring constant is 2000 N/m assuming energy is conserved and friction is neglected, how fast does the cart go when it reached a height of 0.19 m?

2. A person on a bicycle starts from rest. The mass of the bike and rider is 62 kg. If the person uses a constant force of 2.5 N of force to pedal. How fast is the rider going after traveling a distance of 5 km? Is this a reasonable speed? Explain why or why not.

3. For the following scenarios explain who does the most work and who delivers the most power:

   - Two male cheerleaders, Griffin and Hudson, lift their partners up in the air. Both the flyers weigh 110 pounds. Griffin lifts his partner over his head five (approximately 2 ft) times in 5 min. Hudson lifts his partner over his head (the same height) five times in 2 min.
   - The soccer team is running sprints up a hill. The goalie is twice as massive as the smallest striker. The strikers climb the same distance along the hill in half the time of the goalie.

References

1. Arvid Eide, Roland Jenison, Larry Northup, Steven Mickelson (2012), Engineering Fundamentals and Problem Solving (7th Edition), McGraw Hill Education, New York, ISBN-13: 978-0073385914.
2. Henderson, T. (n.d.). The Physics Classroom. Retrieved March 8, 2021, from https://www.physicsclassroom.com/
3. Chisholm, Hugh, ed. (1911). "Coulomb, Charles Augustin". Encyclopædia Britannica. 7 (11th ed.). Cambridge University Press. Retrieved March 15, 2021 from https://en.wikisource.org/wiki/1911_Encyclop%C3%A6dia_Britannica/Coulomb,_Charles_Augustin
4. John Anderson (2017), Fundamentals of Aerodynamics (6th Edition), McGraw Hill Education, New York, NY, ISBN-13: 978-1259129919.
5. Wolfram, Stephen (2002). "A New Kind of Science." Wolfram Media, Champaign, IL, ISBN: 1-57955-008-8.
6. Yildiz, I.; Liu, Y. (2018). "Energy units, conversions, and dimensional analysis". In Dincer, I. (ed.). Comprehensive energy systems. Vol 1: Energy fundamentals. Elsevier. ISBN 9780128149256.

Chapter 18
Mathematics, Models, and Reasoning

Abstract It is important that engineers have a good understanding of mathematics because, as it was stated in Chap. 11, engineering problems are mathematical models of physical situations. Engineers use a wide variety of mathematic concepts when they are formulating and solving problems. Different physical situations have different forms, so this chapter will discuss various mathematical models that are commonly applied, types that are both linear and nonlinear. This chapter will also introduce some common mathematical operations that are regularly used in engineering, such as linear interpolation and solving systems of linear equations. We will also discuss common mathematical symbols and variables used and talk about how to organize and reconstruct word problems into steps and mathematical operations. This is just an introduction. Throughout your engineering curriculum, you will take separate math classes covering topics such as calculus, linear algebra, and ordinary differential equations.

By the end of this chapter, students will learn to:

- Identify mathematical operators, symbols, and variables in both word and symbolic form.
- Define what a mathematical model is and explain a model's purpose.
- Explain the classifications of mathematical models and describe the stages used to fully create a mathematical model.
- Define what a linear model is.
- Explain the characteristics of linear models and give examples of where linear models are used.
- Calculate unknown values using linear models.
- Define what nonlinear power law and polynomial models are.
- Explain the characteristics of power law and polynomial models and identify examples of each.
- Calculate unknown values using power law and polynomial models.
- Define what exponential and logarithmic models are.

© Springer Nature Switzerland AG 2022
M. Blum, *An Inquiry-Based Introduction to Engineering*,
https://doi.org/10.1007/978-3-030-91471-4_18

- Explain the characteristics of exponential and logarithmic models and identify examples of each.
- Explain the relationship between a linear model and a logarithmic model.
- Solve for unknown values using strategies involving linear equations.
- Read word problems and reconstruct them into mathematical statements.

18.1 Mathematical Symbols and Variables

Mathematics is a language. Mathematicians, scientists, and engineers use this language to communicate ideas with logic, precision, and clarity. Like any language, it uses its own terminology and symbols. You have learned much of this language in your previous schooling, learning about arithmetic operational symbols such as plus, minus, division, and multiplication. Over the course of your engineering degree, you will learn many more mathematical symbols, operations, and their meanings. Along with mathematic operations, you will increase the use of the Greek Alphabet. The Greek alphabet is commonly used in math and engineering as variables or to represent important engineering parameters. It is important that you understand what they mean and how to use them properly. This is so that communication will be clear. Whether it be communication between yourself, other students, your professors, or your boss. This next set of questions lets you practice identifying mathematical language in both word and symbolic form.

18.1.1. **Name as many mathematical symbols and operations as you can. Write in their mathematical form.**
18.1.2. **Fill out Table 18.1 with mathematical symbols.**
18.1.3. **Name as many Greek alphabetical symbols as you can. Write in their mathematical form and tell what property they stand for.**
18.1.4. **Fill out Table 18.2 with the Greek Alphabet symbol.**

18.2 Introduction to Mathematical Models

Engineers use mathematical models to make real world problems relatable [1]. We take physical phenomenon, applications, or problems and translate them into mathematic formulations. Math language is precise and concise, and many times computers can be used to perform taxing computations. In engineering, modeling is used to achieve a variety of objectives including

- Provide insight to help develop scientific understanding
- Testing the effects of a change in a system
- To aid in making decisions
- Simulate experiments that may be difficult to create physically
- To visualize and communicate understanding and interpretation of a problem

Table 18.1   Common phrases for mathematical operations

Symbol	Phrase	Symbol	Phrase
	Plus or positive		Absolute value of x
	Minus or negative		Proportional to
	Plus or minus		Therefore
	Multiplication		Summation
	Division		Integral
	Ratio		Factorial, (e.g., $5 \times 4 \times 3 \times 2 \times 1$)
	Less than		Delta indicating difference
	Greater than		Full Differential
	Much less than		Partial Differential
	Much Greater than		Gradient
	Equal to		Pi, value is 3.1415926
	Approximately equal to		Infinity
	Not equal to		Degree
	Identical with		Parentheses
	Equal to or less than		Square Brackets
	Equal to or greater than		Curly Braces

Table 18.2   The Greek Alphabet

Capital letter	Lowercase letter	Name	Capital letter	Lowercase letter	Name
		Alpha			Nu
		Beta			Xi
		Gamma			Omicron
		Delta			Pi
		Epsilon			Rho
		Zeta			Sigma
		Eta			Tau
		Theta			Upsilon
		Iota			Phi
		Kappa			Chi or khi
		Lambda			Psi
		Mu			Omega

The next section of questions helps introduce us to mathematical modeling.

18.2.1. **In your own words, explain what is a model?**

18.2.2. **Name as many components of a mathematical model as you can.**

18.2.3. **In your own words, explain the purpose and importance of mathematical models.**

18.2.4. **Give examples of mathematical models you know of (they could have been discussed in previous chapters or in previous classes).**

In modeling, students need to make choices about what is important, decide what mathematics to apply, and determine whether their solution is useful. The process of developing a mathematical model is formulaic. Specfically, it is useful to divide the process of modeling into discrete activities. There are many different theories about modeling, but essentially four broad categories need to be addressed [1]. Shown in Fig. 18.1 the first phase is **Building** the model. During the building phase, the objectives of the model are clearly defined, mathematical equations are chosen that most closely describe the system (we will learn some standard equations in the subsequent sections), and find any assumptions or simplifications that can be made. The next phase is to **Solve** the model based on the inputs from the modeled system or problem. Analyzing or solving the model can utilize many different mathematical approaches. The third step is to **Study** the model. For this, we work to validate to see how well the model matches the natural behavior we are looking at. Finally, we **Use** the model in the way we need, such as predicting future behavior or visualizing the current behavior of a system. The process is iterative and can be repeated as needed to improve the accuracy or precision of the model.

When in the building stage, there are a variety of types of mathematical models that can be chosen, depending on the objective of analysis. There are two main categories which represent the extreme of modeling. The first category is based on the type of outcome they predict. **Deterministic** models ignore variation so the output of the model is fully determined by the parameter values and the initial conditions. Therefore, if you give it a certain starting point, the model will always give you the same output for that input. On the other hand, **Stochastic** models possess some inherent randomness. The same set of parameter values and initial conditions will lead to a variety of different outputs.

The second category to choose between models deals with the level of understanding on which the model is based. **Mechanistic** modeling uses theory or law that already exists and has been proven or developed. Lots of the modeling you

Fig. 18.1 Main categories and process for creating a mathematical model

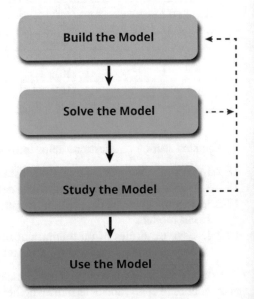

perform in engineering is mechanistic, equations use are already developed theories. **Statistical** modeling is based on having a large set of data and then fitting a model to that. This is often used when underlying theories do not exist. They are more likely to give good predictions, especially where underlying theory does not exist—or known/unknown variables are suspected to be playing a part.

The categories can be used complementary with each other. For example, you can have a mechanistic deterministic model. There are also subclassifications of modeling depending on the applications. Table 18.3 describes different types of models that can all be used within the two broader categories and the reasons for their use. This next set of questions helps us explore creating and identifying models.

18.2.5. **Which method would you most likely use for predicting the outcome of population growth based on a country's census data; mechanistic or statistical?**

18.2.6. **Which method would you most likely use to track planetary motion; mechanistic or statistical?**

18.2.7. **Using Table 18.3, label what type of mathematical model would be applied to examine the following applications.**

- **Force analysis of a truss on a bridge**
- **Fluid flow through a water pipe**
- **Enthusiasm of first year engineering students**
- **Temperature of engineering classroom throughout the day**

Use the scenario for the following questions: A company uses a cardboard shipping box to send purchased items to customers, shown in Fig. 18.2. The size of the box is initially 200 × 300 × 400 mm, and the cardboard is 5 mm thick. Someone suggests that using 4 mm cardboard will allow more items to be packed in the boxes.

18.2.8. **What is the objective of developing a model for this scenario?**

18.2.9. **Choose an accurate mathematical equation to describe the system (make sure all variables are taken into account).**

18.2.10. **Are there any assumptions or simplifications that can be made?**

18.2.11. **Use the model to solve the problem.**

Table 18.3 Different types of modeling methods

Type of model	Uses	Type of model	Uses
Static	Models things in equilibrium state	*Dynamic*	Models things that change with respect to time
Continuous	Models data that can be measured at every value (uses differential equations)	*Discrete*	Models data that can only be measured at certain values (uses difference equations)
Qualitative	Used to make general descriptions about responses (no numbers)	*Quantitative*	Used to make detailed predictions about responses (uses numbers)

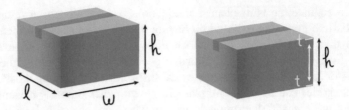

Fig. 18.2 Cardboard shipping box used for Sect. 18.2, showing the variables for the model

18.2.12. **Are there any other issues we might want to consider for the next iteration of the model? (Think about the assumptions that were made)**

The next few sections cover general models that are important to engineers, because the trends of these mathematical functions have been applied within a variety of different engineering disciplines to model subject-specific behavior.

18.3   Linear Model

The simplest form of equations that is commonly used to describe a variety of engineering situations is the linear model. A **linear model** uses an equation where two quantities or variables have a constant rate of change relationship between each other. Visually, linear models are represented by graphing straight lines that are either purely horizontal, purely vertical, or at a constant positive or negative slant. There are three main parameters that need to be identified to create a linear model:

1. **The variables**. One variable will be *independent variable*, that is the variable whose change causes another variable to change. The other variable will be the *dependent variable*. This variable is altered by the change in the independent variable. Many times, the variables are denoted by the symbols x and y, but all through this book there are examples of linear equations that have used other labels, and you will encounter many more other variable labels throughout your engineering curriculum.
2. **The slope**. The slope is the rate of change that the dependent variable experiences as the independent variable changes. The slope is defined as the ratio of the vertical change between two points (also known as the rise) to the horizontal change between the same two points (also known as the run). Typically, the slope is denoted by variable m.
3. **Intercept**. The intercept, often called the initial value, is the constant value of the dependent variable when the independent variable equals zero. It is typically denoted by variable b.

To write a linear model we need to identify the variables, the rate of change, and the initial value. Once we have written a linear model, we can use it to solve all types of problems. This next set of questions guides us to write linear models and solve problems with them.

18.3.1. **What is the general equation for a linear model (you can use symbols _x_ and _y_ for your variables).**

18.3.2. **In the equation you wrote in the previous question, identify the symbol you used for the independent variable.**

18.3.3. **In the equation you wrote in the previous question, identify the symbol you used for the dependent variable.**

18.3.4. **Figure 18.3 represents the position of a car moving at a constant velocity. In Fig. 18.3, what is the independent variable?**

18.3.5. **In Fig. 18.3, what is the dependent variable?**

18.3.6. **In Fig. 18.3, what is the rate of change of the model?**

18.3.7. **In Fig. 18.3, what is the intercept of the model?**

18.3.8. **What is the equation of the linear model that represents this car's position?**

18.3.9. **The engineering rocket club has 30 members and plans to increase membership by five students each year. Write a model that can help the club track the relationship between the number of members over each year the club operates. Be sure to identify all the parameters of the model!**

18.3.10. **The position of a car (in kilometers) at a time (in hours) is given by the following equation: $p = 75t + 200$. How far does the car travel in 2 h?**

18.3.11. **Draw a graph (make sure to label axes and important parameters) of the following linear models:**

    1. $y = 4x + 6$
    2. $y = 5$
    3. $x = 7$

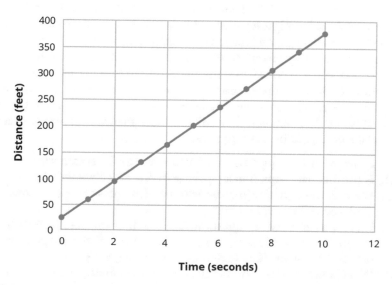

Fig. 18.3 Graph representing the position of a car moving at a constant velocity

4. $y = 10 - 3x$

18.3.12. **Identify three linear models that have been introduced in other chapters of this book and write down the equations.**

18.4   Power Law and Polynomial Models

While linear models are simple and straightforward to use, most engineering relationships between dependent and independent variables are nonlinear. Therefore, other mathematical equations are used to better capture or predict natural behavior. Another group of models are power law and polynomial models. A **power model** consists of an equation where a variable (typically the independent variable) is raised to a number power. Power models are represented by equations with the general form $y = kx^p$. Where x and y again represent the independent and dependent variable, k represents a coefficient (typically it is a constant value), and p represents an exponent value. The exponent is a fixed real number that raises the variable to a power. This tells you how many times to multiply that number by itself. For example, $2^4 = 2 \times 2 \times 2 \times 2 = 16$.

A **polynomial** model is an equation which consists of the sum of a finite number of variable terms. Now the independent variable term appears more than once. Each term is the product of a number and a variable raised to a nonnegative integer power. Polynomial models are represented by equations with the general form $y = C_n x^n + \cdots + C_2 x^2 + C_1 x^1 + C_0 x^0$. Where x and y again represent the variables, each C_i represents a coefficient (typically it is a constant value) and can be any real number, and x^i terms are the variable raised to a nonnegative integer exponent. The *degree of a polynomial function* is the highest power of the variable that occurs in a polynomial. The term containing the highest power of the variable is called the *leading term*. The coefficient of the leading term is called the *leading coefficient*.

Throughout your engineering curriculum, you will solve engineering problems with power law and polynomial models. You will learn to understand how to graph power law and polynomial functions and to find their zeros (roots). This is important because the graph shapes and root values will have physical meaning in engineering situations. For this course, we will only learn to identify power law models, polynomial models, and their components.

18.4.1. **Is the following equation a power law: $A = \pi r^2$? Explain why.**
18.4.2. **Is the following equation a power law: $A = \pi 2^r$? Explain why.**
18.4.3. **What is the equation for the volume of a sphere? Is it a power law model or polynomial model?**
18.4.4. **Based on the previous question, label each of the components for the equation of the volume of a sphere with each of the parameters that make up its model (i.e., variables, coefficients, etc.).**
18.4.5. **What is the leading term for the given polynomial: $y = 5x^3 + 2x^2 + 5y$**

18.4.6. **Based on the polynomial in the previous question, what is the coefficient on the leading term?**

18.4.7. **Based on the polynomial in question 18.4.5, what is the leading term's degree?**

18.4.8. **Based on the polynomial in question 18.4.5, what is the degree (order) of the polynomial?**

18.4.9. **Finish fill in the blanks in Table 18.4.**

18.5 Exponential and Logarithmic Models

The final group of mathematical models that are seen often in engineering is exponential and logarithmic models. **Exponential models** use mathematical functions where the independent variable is the exponent. Instead of a power law such as

Table 18.4 Examples of polynomial and power law models used in engineering

Model	Equation	Power law or polynomial
Volume of a cone	V = volume r = radius of cone h = height of cone	
Kinetic energy equation	KE = kinetic energy m = mass v = velocity	
Stopping sight distance	$$S = \dfrac{1}{2g\mu \pm G}V^2 + tV$$ S = stopping sight distance V = initial speed g = gravitational acceleration constant μ = coefficient of friction G = grade of road constant t = driver reaction time constant	
Ideal Gas Law at a constant temperature	$$V = \dfrac{nRT}{P}$$ V = volume P = pressure T = temperature n = quantity of gas constant	
Position of a particle with constant acceleration	$$x = x_0 + v_0 t + \dfrac{1}{2}at^2$$ x = distance (x_0 is initial position) t = time v_0 = initial velocity a = constant acceleration value	

$y = x^2$, an exponential model would be $y = 2^x$. Where the value 2 is the *base*, and the variable in the superscript x is the *exponent*. In engineering and science, exponential functions are commonly used to describe growth and decay models. For example, the previous function $y = 2^x$ would be used to model a growth system, where the value of y doubles every time, you increase the value x by one. A graph of this function is shown in Fig. 18.4. Contrary to that, a decay function shows the value of y decreases every time you increase the x-value. Although the base parameter could be numerous integers, calculus found that a certain value occurred so frequently when modeling growth or decay that it has become the constant for the base of exponential models. Known as the **exponential constant**, Euler's number, or Napier's constant, is an irrational number with an approximate value of $e = 2.718281...$ [2]. Fixing the base, an exponential model form can be written as $y = ce^{kx}$, where $e = 2.718281$, c is a constant that scales the size of the function, and k is a constant that makes the function be growth or decay and it can speed up or slow down the decay/growth. The constant k must make or keep the exponent dimensionless, so it is important to understand the units of k.

As previously explained, the exponent of a function is the superscript above an integer, and it describes how many times to use that number in multiplication. Going back to our example of $2^4 = 2 \times 2 \times 2 \times 2 = 16$. In this example 2 is the base and 4 is the exponent and 16 is the answer. Essentially it is saying that 2 is used 4 times in multiplication to get the number 16. A logarithm is the reciprocal of that. A **Logarithm** says how many of one number to multiply to get another number. It answers the question "what exponent produced this?" Looking at our example once again, except now the logarithm is saying that 2 makes 16 when used 4 times in multiplication. Figure 18.5 shows our example in exponent and logarithm form.

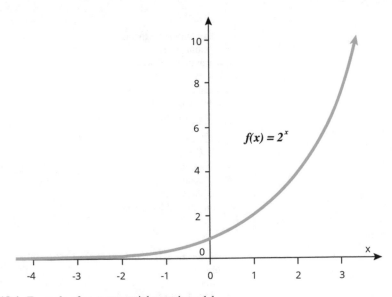

Fig. 18.4 Example of an exponential growth model

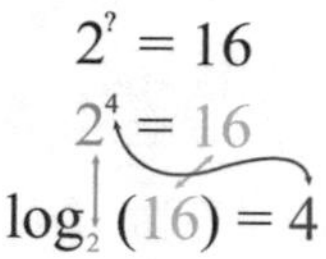

Fig. 18.5 Form of exponential and logarithm examples

$$2^? = 16$$
$$2^4 = 16$$
$$\log_2(16) = 4$$

Table 18.4 Common exponent and log mathematical relationships

Exponent relationships	Logarithmic relationships
$x^n x^m = x^{n+m}$	$\log yx = \log x + \log y$
$(xy)^n = x^n y^n$	
$(x^n)^m = x^{nm}$	
$x^{-n} = \dfrac{1}{x^n}$	$\log \dfrac{x}{y} = \log x - \log y$
$\dfrac{x^n}{x^m} = x^{n-m}$	
$\left(\dfrac{x}{y}\right)^n = \dfrac{x^n}{y^n}$	$\log x^n = n \log x$
$x^0 = 1 \qquad (x \neq 0)$	

Table 18.5 lists common exponent and log mathematical relationships which you will learn about and use throughout your engineering curriculum. Below are some important notes about several logarithmic functions that are defined for their ease of computation, and their relationship with exponentials.

- *The logarithm is the inverse function of exponentiation.* Therefore, they cancel each other out. For example, $\log_a(a^x) = x$ and $a^{\log_a(x)} = x$.
- The *common logarithm* (known as *log*) has a base 10. Meaning its exponent form is $10^x = y$, and its log form is $\log y = x$.
- The natural log (known as *ln*) has a base $e = 2.718281$. Meaning its exponent form is $e^x = y$, and its log form is $\ln y = x$.

Finally, logarithmic scales are used when there is a large range of quantities. It is based on orders of magnitude, rather than a standard linear scale, so the value represented by each equidistant mark on the scale is the value at the previous mark multiplied by a constant. It is important to remember that even though the graph appears to be a straight line it is not linear because the values on the axes are in powers of ten. Therefore, the spacing between each power is scaled to that particular power. It is useful to understand how a log-log scale translates to a linear scale and vise versa because a variety of engineering data will be displayed using Log-Log plots. The next set of questions helps us practice getting familiar with exponential and logarithmic models.

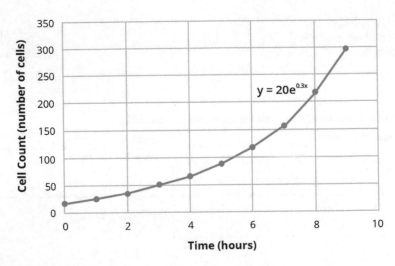

Fig. 18.6 Graph of cell growth for tissue mechanic experiment

18.5.1. If given an equation $y = ce^{kx}$ and the value of k is positive, is the model of growth, decay or constant?

18.5.2. If given an equation $y = ce^{kx}$ and the value of k is negative, is the model of growth, decay or constant?

18.5.3. If given an equation $y = ce^{kx}$ and the value of $k = 0$, is the model of growth, decay or constant?

18.5.4. A biomedical engineer is growing cells in a petri dish for a tissue mechanics experiment. The graph in Fig. 18.6 is the results after several hours of counting cells. Is the model of growth, decay or constant?

18.5.5. Based on the graph and equation in Fig. 18.6, identify the dependent and independent variable.

18.5.6. Based on the graph and equation in Fig. 18.6, what is the initial number of cells that are put in the petri dish?

18.5.7. Based on the graph and equation in Fig. 18.6, what is the growth constant of the cells?

18.5.8. Based on the graph and equation in Fig. 18.6, how long will it take for the cell culture population to grow to 20,000 cells?

18.5.9. Name an example of an exponential (growth or decay) model that you are familiar with (it could be from current or previous classes).

18.5.10. What is the x in the following equation: $\log_3(x) = 6$.

18.5.11. Calculate y in the following equation: $y = \log_5\left(\dfrac{1}{5}\right)$

18.5.12. Based on the Log-Log Scale Figure shown in Fig. 18.7 convert the y-axis in terms into fully written out orders of magnitude.

18.5.13. The generic equation for a log-log line is given in Fig. 18.7 develop a method using two points on the graph $(10^2, 10^0)$ and $(10^3, 10^{-2})$ to find the full equation for the log-log line.

18.5.14. For the graph in Fig. 18.7 develop the full equation for the line by solving for the constants m and b.

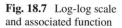

Fig. 18.7 Log-log scale and associated function

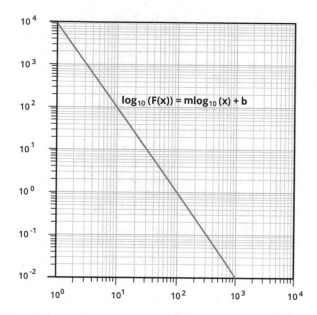

$$\log_{10}(F(x)) = m\log_{10}(x) + b$$

18.6   Useful Math Strategies Using Linear Models

As discussed in Sect. 18.3, many engineering situations can be described or estimated using linear models. In this next section, we will discuss some important standard strategies and procedures for using linear equations to solve problems.

1. Solving systems of multiple linear equations

    Many times, to solve an engineering problem we need to create sets of linear equations that are related by the same unknown variables. A **system of two linear equations** typically involves two equations with the same two variables. There are three strategies to solving a system of linear equations:

    • **Graphing Approach**: Using a graph is a useful strategy because it sometimes helps to visualize the solution. The first step is to solve both equations for the same variable. Then graph both equations on the same Cartesian coordinate system and find the point where the two lines cross. The value of the intersection point is the answer.

    • **Substitution Approach**: The substitution approach is useful when you are only looking for one of the two variables in the equations. The first step is to get the unnecessary variable by itself in one of the equations. Then, take that expression and plug it into the other equation. This removes the unnecessary variable and puts one equation completely in terms of the wanted variable. Finally, solve for the wanted variable using one equation. Also, if you wanted you could now go back and solve the other variable.

    • **Elimination Approach**: For the elimination approach, first rearrange both equations so that when they are written vertically, one on top of the other, the same variable in each equation is aligned. Multiply one (or both) equations by a constant that will allow one of the variables to cancel when the equations are added or subtracted together. Add or subtract the equations and solve for the remaining variable. Also, if you wanted you could now go back and solve the other variable.

The next set of questions give us practice solving systems of two linear equations.

18.6.1. **Using the equations in Fig. 18.8 in their current form, which approach would you use to solve for *x* and *y*; substitution or elimination? Explain why you would use that strategy?**

18.6.2. **Using the equations in Fig. 18.8, solve for both variables using the approach you choose.**

18.6.3. **Using the equations in Fig. 18.8, show your work to solve for the variables using the approach not chosen in the previous question.**

18.6.4. **Use the graphing strategy, labeling axes, and important values, to show the solution to the equations in Fig. 18.8.**

When you have three or more linear equations with variables the strategies become more complex. Solving many linear equations simultaneously is a branch of mathematics known as Linear Algebra. You will be introduced to linear algebra in a mathematics class and practice it throughout your engineering curriculum.

2. Linear Interpolation

   In many engineering classes you will need to find data and look up values in a table. Occasionally, a table will not have the exact increment you need. In order to calculate the value you need, linear interpolation can be used. **Linear Interpolation** is a mathematical method to approximate the value from a data table. Within a table of data, the function is only specified at a limited number or discrete set of independent variable values. Whereas when a function is defined (like that of a straight line) the solution is defined along a continuum. In Linear interpolation, you use the equation of a line to find a new data point, based on an existing set of data points from a table.

   For example, as shown in Fig. 18.9 if we know two points on a line (point 1 (x_1, y_1), point 2 (x_2, y_2)). We want to find the values for point P (x_P, y_P). Since all three points are on the same line, we know the slope *m* is equal to the rise over run between each of the points so

$$m = \frac{\Delta y_{P \to 1}}{\Delta x_{P \to 1}} = \frac{\Delta y_{2 \to 1}}{\Delta x_{2 \to 1}}$$

Fig. 18.8 A system of two
linear equations

$$Eq\ (1) \quad x = 6 - y$$
$$Eq\ (2) \quad 2x + 5y = 6$$

$$m = \frac{y_P - y_1}{x_P - x_1} = \frac{y_2 - y_1}{x_2 - x_1}$$

If we set one of the values for point P (x_P or y_P) we can solve the ratio equation for the other point. This technique is frequently used to approximate nonlinear data as well. The next set of questions helps us practice linear interpolation.

18.6.5. **Consider the graph in Fig. 18.9a which is associated with the table in Fig. 18.9b. Is this model linear?**

18.6.6. **Consider the graph in Fig. 18.9a if we wanted to know the stopping distance a car needed if it was traveling at 53 mph could we read it directly off the table?**

18.6.7. Based on the previous question, what would be your point P values (which is known and which is unknown)?

18.6.8. **Consider the table in Fig. 18.9b, identify which two points would be the known values for your straight-line equation? Identify the x and y-values for each. (Hint: they should be above and below your known point P value)**

18.6.9. **From questions 18.6.6–18.6.8, Find the stopping distance a car needed if it was traveling at 53 mph.**

18.6.10. **Is the answer above reasonable? How do you know?**

18.6.11. **If a car needs a stopping distance of 80 ft before a stop sign, what should be the posted speed limit in miles per hour?**

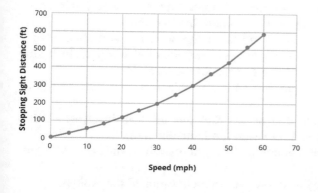

Speed (mph)	Speed (ft/s)	Stopping Sight Distance (ft)
0	0.0	0
5	7.3	21
10	14.7	47
15	22.0	78
20	29.3	114
25	36.7	155
30	44.0	201
35	51.3	252
40	58.7	309
45	66.0	370
50	73.3	436
55	80.7	508
60	88.0	584

(a)   (b)

Fig. 18.9 Model from the American Association of State Highway Officials (AASHO) that estimates the distance a driver needs to stop their car traveling at a certain speed if they detect a risk [4]. (a) Graph of Data, (b) table of data

18.7   Turning Words into Math

One of the most important skills an engineer can acquire is the ability to read or observe a physical problem and to organize and reconstruct it into steps and mathematical operations. This requires visualization and planar spatial reasoning skills. Below are some tips for turning words into mathematic statements.

- Find and circle all the numerical values or symbols. Choose variable symbols if necessary.
- Identify "math language" and key action words (Think back to Sect. 18.1).
- Draw a picture to visualize if necessary.
- Rewrite the main question of the problem in your own words or equations.
- Identify the information you have and what you need to solve.
- Evaluate the steps needed to be taken to solve the main question. There may be supplementary equations and variables that need to be solved.
- Show all the steps as you solve the problem and always check that your final answer makes sense.

The next set of questions helps us grow the skill of turning word problems into solvable math statements.

18.7.1.   **Solve the following math problem for the variable x; $x^2 - 5x - 6 = 0$.**

For the next set of questions look at the following scenario:
You have a square pool. You want to replace it with a rectangular pool. One side of the new pool increases the length by 2 ft, and the other side increases the length by 3 ft. This doubles the area of the pool. What is the size of the original pool?

18.7.2.   **Circle all the numerical values and variables in the scenario.**
18.7.3.   **What is the variable you need to choose yourself? Namely What variable do you need a symbol for? (Call that variable x)**
18.7.4.   **Identify "math language" and key action words by underling them.**
18.7.5.   **Draw a picture of the original pool and then the new pool, including the variable and the changes stated in the scenario.**
18.7.6.   **Rewrite the main question of the problem in your own words or equations.**
18.7.7.   **What are the main dimensions that this problem involves? (Think back to Chap. 10 and the seven fundamental ones).**
18.7.8.   **What supplemental equation that you already know deals with this problem?**
18.7.9.   **Develop a strategy to answer the main question of the problem.**
18.7.10. **Answer the main question of the problem.**
18.7.11. **What was the original area of the pool? What is the new area of the pool?**
18.7.12. **Explain the relationship between question 1 of this problem set and the solution to the scenario.**

End of Chapter Problems

IBL Problems

IBL1: The following table shows the cost of a slice of pizza dependent on the number of extra toppings. Answer the following questions:

Toppings (x)	Cost (y)
2	$3.50
3	$4.25
5	$5.75

1. Based on the table, how much does it cost to have five toppings on a slice of pizza?
2. Based on the table, how much does it cost to have three toppings on a slice of pizza?
3. Based on the previous two questions, how much does it cost to add two additional toppings on a slice of pizza?
4. Based on the table, how much does it cost to have two toppings on a slice of pizza?
5. Based on the previous two questions, how much does it cost to add one additional topping on a slice of pizza?
6. How much does a slice of pizza cost with four toppings on it?
7. How much does a slice of plain cheese pizza cost (only one topping)?
8. Explain the relationship between the cost of a slice of pizza and the number of extra toppings.
9. Develop an equation for the relationship between the cost of a slice of pizza and the number of extra toppings.

IBL2: You are throwing a Halloween party, so you go to the store to get a mix of candy. Chocolate candy costs $1.20 per pound and sour candy costs $1.50 per pound. You want to buy 15 pounds of sour candy. Answer the following questions

1. How much does one pound of chocolate candy cost?
2. How much does two pounds of chocolate candy cost?
3. How much does three pounds of chocolate candy cost?
4. If the amount of chocolate candy is labeled as X, write an equation for the cost of X amount of chocolate candy.
5. How much does one pound of sour candy cost?
6. How much do 15 pounds of sour candy cost?
7. From the previous questions, develop an equation for the total cost of a mixture of an unknown amount of chocolate candy and 15 pounds of sour candy.
8. Develop an equation for the total weight of a mixture of an unknown amount of chocolate candy and 15 pounds of sour candy.

9. If you want a mixture of 15 pounds of sour candy and chocolate candy that is
    $1.30 per pound, explain a procedure for how you would determine the amount
    of chocolate candy you need to buy?
10. Determine the amount of chocolate candy you need to buy if you want a mix-
    ture with 15 pounds of sour candy that is $1.30 per pound.

Practice Problems

1. The US Census Bureau collects information about many aspects of the US population [3]. Below is a Table showing the Median Age of Males and Females at
   First Marriage by Year. Answer the following questions:

   - Estimate the median age for the first marriage of a male and female in the
     year 1915.
   - Estimate the median age for the first marriage of a male and female in the
     year 1956.
   - Estimate the median age for the first marriage of a male and female in the
     year 1979.
   - Estimate the median age for the first marriage of a male and female in the
     year 2002.

Year	Average Age of Males at First Marriage	Average Age of Females at First Marriage
2020	30.5	28.1
2000	26.8	25.1
1980	24.7	22.0
1960	22.8	20.3
1940	24.3	21.5
1920	24.6	21.2
1900	25.9	21.9
1890	26.1	22.0

2. What are the roots of the following polynomials

   - $y = 3x^2 - 5x - 4$
   - $y = x^2 - 14x + 49$
   - $y = x^2 + 2x - 15$

3. For the following graph, explain (i) what type of model it is and its characteristics, (ii) state the mathematical formula for the model, and (iii) name the constants associated with the model, their values, their units, and physical meanings.

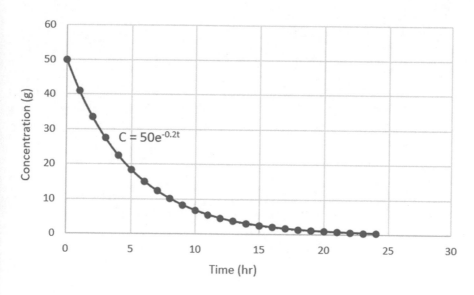

$C = 50e^{-0.2t}$

References

1. Marion, G., & Lawson, D. (2015). An Introduction to Mathematical Modelling.
2. Weisstein, Eric W. "e". mathworld.wolfram.com. Retrieved May 5, 2021.
3. Estimated Median Age at First Marriage by Sex: 1980 to the Present. United States Census Bureau. https://www.census.gov/data/tables/time-series/demo/families/marital.html
4. Abdulhafedh, A. (2020). Highway Stopping Sight Distance, Decision Sight Distance, and Passing Sight Distance Based on AASHTO Models. Open Access Library Journal, 7, 1–24. https://doi.org/10.4236/oalib.1106095

Index

Printed in the United States
by Baker & Taylor Publisher Services